T0076340

Gnotobiotic Mouse Technology

AN ILLUSTRATED GUIDE

Gnotobiotic Mouse Technology

AN ILLUSTRATED GUIDE

Chriss J. Vowles • Natalie E. Anderson
Kathryn A. Eaton

CRC Press
Taylor & Francis Group
Boca Raton   London   New York

CRC Press is an imprint of the
Taylor & Francis Group, an **informa** business

CRC Press
Taylor & Francis Group
6000 Broken Sound Parkway NW, Suite 300
Boca Raton, FL 33487-2742

© 2016 by Taylor & Francis Group, LLC
CRC Press is an imprint of Taylor & Francis Group, an Informa business

No claim to original U.S. Government works

Printed on acid-free paper
Version Date: 20150917

International Standard Book Number-13: 978-1-4987-3632-9 (Paperback)

This book contains information obtained from authentic and highly regarded sources. Reasonable efforts have been made to publish reliable data and information, but the author and publisher cannot assume responsibility for the validity of all materials or the consequences of their use. The authors and publishers have attempted to trace the copyright holders of all material reproduced in this publication and apologize to copyright holders if permission to publish in this form has not been obtained. If any copyright material has not been acknowledged please write and let us know so we may rectify in any future reprint.

Except as permitted under U.S. Copyright Law, no part of this book may be reprinted, reproduced, transmitted, or utilized in any form by any electronic, mechanical, or other means, now known or hereafter invented, including photocopying, microfilming, and recording, or in any information storage or retrieval system, without written permission from the publishers.

For permission to photocopy or use material electronically from this work, please access www.copyright.com (http://www.copyright.com/) or contact the Copyright Clearance Center, Inc. (CCC), 222 Rosewood Drive, Danvers, MA 01923, 978-750-8400. CCC is a not-for-profit organization that provides licenses and registration for a variety of users. For organizations that have been granted a photocopy license by the CCC, a separate system of payment has been arranged.

Trademark Notice: Product or corporate names may be trademarks or registered trademarks, and are used only for identification and explanation without intent to infringe.

Visit the Taylor & Francis Web site at
http://www.taylorandfrancis.com

and the CRC Press Web site at
http://www.crcpress.com

This work is dedicated to Gayle Francis:
helper, friend, and colleague.

Without you, this would not have been possible.

Contents

Preface and Acknowledgments

No instructional manual can possibly be completed without the help and support of innumerable individuals contributing suggestions, ideas, and encouragement. For this, we particularly thank the investigators who use and support the University of Michigan Germ Free Mouse Resource and our many colleagues and friends who have participated in the establishment and support of our resource over the years. Particular thanks goes to Trenton Schoeb for establishing and maintaining the gnotobiology e-mail list, which has become a major avenue for communication between laboratories throughout the world and which we recommend to anyone interested in gnotobiotic technology (see Chapter 18 for more information). Particular thanks also to Howard Rush, who initiated the Michigan Germ Free Laboratory and provided moral as well as financial support during the early years.

In addition to the moral support supplied by these individuals, we would like to acknowledge the following people for their help and support. Clinton Fontaine and Sara Poe assisted by supplying information and by critical reading of the text, as well as by patiently tolerating our frequent disappearances for the purpose of writing this manual. Bob Dysko, Harry Mobley, and Vincent Young provided encouragement and financial support for the Germ Free Resource, as well as management suggestions and moral support. Gabriel Nunez and Grace Chen provided much encouragement as well as helpful management suggestions. And of course, we thank our families for putting up with us during the writing process.

About the Authors

Chriss Vowles comanages a multiinvestigator germ-free research laboratory at the University of Michigan. He began his career at the University of Michigan in 2003, working full time as a husbandry technician in the Unit for Laboratory Animal Medicine. In 2006, he discovered gnotobiotic technology. At that time, the Germ Free Laboratory was just starting. Chriss joined the research group on the ground floor and has been growing with it ever since. The idea of maintaining a germ-free, complex organism fascinated him then and still captivates him today. After 8 years of maintaining a germ-free colony of mice, the laboratory is still evolving, he is still learning, and his trials, tribulations, and rewards are constant.

Natalie Anderson is currently a research technician lead at the University of Michigan's Germ Free Core. In 2009, she graduated from Michigan State University with a bachelor's degree in animal science and accepted a position with SoBran's contract for the National Institute of Allergy and Infectious Diseases (NIAID). In December 2010, she returned to Michigan to take a position with the University of Michigan's Unit for Lab Animal Medicine. Shortly thereafter, she started working part time in the Germ Free Core and quickly learned to love the daily challenges involved in maintaining a germ-free colony. Natalie joined the core full time in July 2013.

Kathryn Eaton has been working with germ-free and gnotobiotic animal models for 30 years. She started her gnotobiology career at Ohio State University, where she did her PhD research on *Helicobacter pylori* in gnotobiotic piglets. She went on to study mice several years later, and in 2004 she established the University of Michigan

Germ Free Mouse Laboratory, which she now directs. Dr. Eaton is a board-certified veterinary pathologist with research interests in bacterial enteric disease and immunology. In addition to *H. pylori*, she has worked with gnotobiotic animal models of shigatoxigenic *Escherichia coli*, inflammatory bowel disease, and most recently, the roles of the enteric microbiome in health and disease. She has long been a staunch supporter of gnotobiotic research and is greatly encouraged by the recent explosion of new investigations, methods, and models that have enhanced the utility and availability of germ-free animal models of disease.

The authors can be reached through the University of Michigan Germ Free Mouse Core at http://research.med.umich.edu/ulam-germ-free

A Brief History
of Germ-Free Life

Germ-free life was first contemplated as early as 1885, when Louis Pasteur broached the question of whether animals could live in the absence of microbes (see Reyniers, Trexler, and Ervin 1946). Over the next 50 years or so, several attempts were made to establish germ-free animals of various species (reviewed in Reyniers, Trexler, and Ervin 1946). The first in-depth studies of germ-free animals, however, did not occur until about 1939, when James Reyniers at the Laboratories of Bacteriology, University of Notre Dame (LOBUND), established in 1928, successfully derived and reared germ-free rats.[1] In 1946, Reyniers observed: "The animal is thought of as a single biological system whereas it is actually a complex consisting of two systems, the microbial and the animal *per se*" (p. 88). His research interest was understanding the interactions between animal and microbe, particularly concerning nutrition, and his work formed the basis for modern germ-free technology. The laboratory published three volumes of LOBUND Reports, as well as several other publications,[1-6] describing their methods for derivation and maintenance of germ-free animals and studies of the nutritional requirements of germ-free rats and chickens. These reports are, in fact, an excellent resource for information on the history of germ-free animals as well as germ-free methodology, and many of the methods described are similar to those used today.

At LOBUND, investigators derived and maintained many species, including germ-free guinea pigs, chickens, monkeys, cats, dogs, rabbits, mice, flies, and fish.[1] The first germ-free rats were derived at LOBUND by cesarean section and were hand raised in isolators by round-the-clock feeding using hand-drawn glass pipettes as feeding tubes. Cesarean section and hand raising are still used to produce germ-free piglets, puppies, kittens, ferrets, calves, and other larger animals, but most laboratories that use germ-free animals today primarily use mice and breed them in the germ-free environment. Because mice and rats can live their entire lives germ-free, rederivation is necessary only when new strains are to be introduced, and most laboratories now use embryo transfer to produce new germ-free mouse strains, obviating much of the manual labor necessary to raise neonates inside the germ-free isolator.

In addition to the considerable research done on derivation, nutrition, physiology, and management of germ-free animals, LOBUND was the source of the modern bubble-type isolator, still the most common type of biocontainment used today to breed and house germ-free animals. Development of the "Trexler isolator," designed by Philip Trexler,[2] was intended to facilitate widespread availability of germ-free technology among other laboratories of bacteriology; in fact, it succeeded in doing just that. Prior to the Trexler isolator, germ-free animals were kept in what were essentially self-contained autoclaves, with solid steel sides, steel air locks, and only small viewing windows and limited access ports. Today's isolators, with clear plastic canopies, flexible gloves, positive-pressure airflow, and easily manipulated double air locks markedly facilitate animal husbandry as well as experimental techniques.

That said, there can be no doubt that keeping animals germ-free still involves huge challenges and requires considerable labor and vigilance (not to mention expense). Bacteria are everywhere and are remarkably adept at colonizing surfaces, particularly in the absence of competition. Normal animals, as so aptly observed by Reyniers, are in essence a complex of animal and bacterial systems, which will revert to their normal multispecies condition at the first opportunity. Thus, obtaining and maintaining germ-free animals is complex and requires knowledge, training, and meticulous attention to detail. Consequently, the current manual is intended as an illustrated step-by-step guide to modern methods used to produce, maintain, and use germ-free mice in a research support facility.

In addition to providing a guide to modern methods in gnotobiology, the second justification for this manual relates to current trends in biomedical research. Over the past 20 years, the mouse has become the principal laboratory animal species used in basic research on physiology and disease. Manipulated mouse models and genetically engineered mice have become increasingly common. More

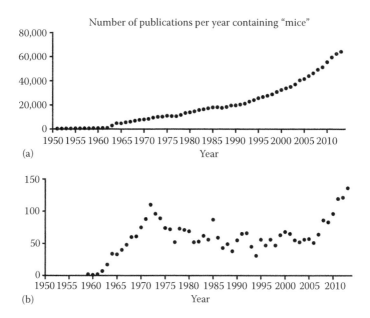

Figure 1.1 Results of a PubMed search for title or abstract containing either (a) "mice" or (b) "germ-free, germ-free, or germfree mice."

recently, interest in studying the microbial inhabitants of vertebrate animals has also begun to grow. A search of PubMed revealed that while the number of publications relating to research using mice in general has risen steadily over the past half century, use of germ-free mice peaked in the early 1970s, declined, and then began to increase again (see Figure 1.1). In the last 8–10 years, the number of publications that referred to germ-free mice has risen precipitously, emphasizing the need for a user manual on establishing and running a new germ-free mouse facility. More institutions are beginning to set up germ-free mouse colonies, and easily accessible information is essential for the success of these ventures. The methods and technologies necessary to successfully establish and run a germ-free mouse colony are not new, but they can appear mysterious and unreachable, and particularly for the neophyte, efforts can be frustrating in the absence of clear instruction and advice.

Over the past few years, we have had many requests for advice and information, and it has become clear that a manual such as this was needed to assist new investigators setting up their germ-free mouse facilities and help them avoid the errors and frustrations that we have experienced in establishing our germ-free mouse core. Our goal is to present the methods that we have found to be successful and the sources and types of equipment and supplies that we have found useful. We hope that these methods and their adaptations will be helpful for those setting up a new germ-free mouse facility as well as those with established facilities interested in alternative techniques.

References

1. Reyniers J, Trexler PC, Ervin RF. Rearing germ-free albino rats. In: Reyniers JA, Ervin RF, Gordon HA, eds. *Lobund Reports*. Volume 1. Notre Dame, IN: University of Notre Dame Press, 1946:1–87.

2. Reyniers J, Trexler PC, Ervin RF. Germ-free life applied to nutrition studies. In: Reyniers J, Ervin RF, Gordon HA, eds. *Lobund Reports*. Volume 1. Notre Dame, Indiana: University of Notre Dame Press, 1946.

3. Reyniers JAR, Gordon HAE, Robert F, Ervin RFG, Helmut A, Wagner MW, Morris. *Germ Free Life Studies* no. 3. Volume 3. Notre Dame, IN: University of Notre Dame Press, 1960:1–203.

4. Reyniers JAR, Trexler PC, Kopac MJ, Hildebrand EM, Woolpert OC, Hudson NP, Glaser RW, White PR, Rosenstern I, Kammerling F, Wells WF. *Micrugical and Germ-Free Methods*. Springfield, IL: Charles C. Thomas, 1943.

5. Reyniers JR, Ervin RF, Gordon HA. *Lobund Reports: Germ free Life Studies*. Volume 2. Notre Dame, IN: University of Notre Dame Press, 1949:1–149.

6. Reyniers JA, Trexler PC. The germ-free technique and its application to rearing animals free form contamination. In: Reyniers JA, ed. *Micrugical and Germ-Free Methods*. Springfield, IL: Charles C. Thomas, 1943:114–143.

7. Trexler PC, Reynolds LI. Flexible film apparatus for the rearing and use of germfree animals. *Applied Microbiology* 1957; 5:406–412.

Overview of
Gnotobiotic Technology

What Are Germ-Free Animals?

In today's world, laboratory animals, particularly rodents, are housed in what are called *biological barriers*. Biological barriers, of which germ-free is the most strict, are designed to keep animals free from contact with microbes, and each level of barrier is defined by the microbes that are excluded and by the methods needed to create and preserve that particular barrier. Definitions and descriptions of the various levels of microbial containment are as follows:

Conventional animals are kept in open cages with no biological barrier and are not screened for microbes. Conventional housing is rare in modern animal facilities.

Specific pathogen free (SPF) animals are free of specific, named pathogens. They are derived by various methods, such as cross fostering, cesarean rederivation, or antibiotic or anthelminthic treatment. SPF animals are housed in HEPA-filtered cages, handled in laminar flow hoods, and screened for specific bacterial, viral, fungal, and protozoal pathogens, depending on the institution. Most mice used in research today are SPF, although the pathogens that are eliminated may or may not be specified. These mice have a full complement of nonpathogenic microbiota, which is generally not defined and varies markedly depending on the source, level of biocontainment, and institution, among other things.

Gnotobiotic animals are animals with a known, defined microbiota. Gnotobiotic mice are usually obtained by starting with germ-free mice and colonizing them with one or more microbes. They may be kept in HEPA-filtered cages or in germ-free isolators, but they must be monitored to ensure that they continue to harbor the desired microorganism(s), and that no unintended contamination occurs. Many commercially available SPF mice were originally produced by colonizing germ-free mice with a known cocktail of normal enteric microbes.

Germ-free (also referred to as germ-free, germfree, or axenic) animals contain no detectable microbial life. They are considered free of all bacteria, fungi, parasites, and viruses (with the exception of endogenous viruses), and they must be kept in double air-locked, HEPA-filtered isolators that are maintained with strict asepsis.

"Conventionalized" animals are those that started germ-free but were removed from the isolator and exposed to normal microbiota either by contact or by administration of (usually) intestinal contents or feces of SPF mice. This term is not precisely correct because animals given SPF feces are more like SPF than conventional animals. Nevertheless, the term is in common use, and its meaning is generally understood.

How Do We Get Germ-Free Mice?

As mentioned, the first germ-free rodents were derived by a laborious process of cesarean section and hand raising. The level of difficulty of this procedure was well documented by Reyniers et al.[1] and is worth reading. Once germ-free rodent colonies were established, however, it became possible to derive new germ-free strains by cesarean section, aseptic transfer of the pregnant uterus into a germ-free isolator, and fostering of the offspring on a germ-free dam (see Chapter 14). While this was an improvement over hand raising, it still involved date mating donor and foster dams, surgically removing pups, passing the pups through a disinfectant bath, and ensuring that the foster dam accepted and fed them. Neonatal death was common because of immaturity, complications of transfer into the isolator, maternal neglect, and so on. Today, the most common method of generating new germ-free mouse strains is embryo transfer. Donor and foster dams still must be date mated, and surgery is still necessary to provide vasectomized males and to collect and transfer embryos, but maternal acceptance is not a problem, and contamination is easier to avoid.

For new facilities starting up, the best way to obtain mice is from an existing facility. Most commonly used strains can be obtained germ-free. If a mouse strain is not available germ-free, however, it must be rederived either by cesarean section or by embryo transfer.

Germ-Free Housing and Barrier Types

Traditionally, germ-free mice are kept in Trexler-type, soft-sided, "bubble" isolators.[2] These vary considerably in detail but are all based on the same principle. They consist of a plastic bubble (the canopy) with at least one double air lock (the port) and at least one pair of gloves (Figure 2.1). Isolators are maintained under positive air pressure by pumps that blow air in through a HEPA filter, and air moves passively out of the isolator through a second filter. The isolator contains mice, food, water, bedding, and any supplies needed for husbandry or experimental procedures. All supplies are sterilized prior to entry into the isolator, and all entries and exits are performed aseptically via the double air lock to maximize protection from inadvertent contamination. Isolators can be constructed in various sizes for housing as few as 1 or 2 cages or as many as 50 cages. All must be strictly monitored for sterility, however, and prevention of contamination is complex. In addition, if an experiment involves gnotobiotic mice, a single isolator can only house a single colonized group. This is because any organism that is present in the isolator will rapidly spread to all cages in that isolator, prohibiting bacteriologic isolation of groups within the isolator. For that reason, each experimental group requires its own isolator, and once the experiment is complete, the isolator must be completely decommissioned and resterilized before reuse.

In our laboratory, in addition to bubble isolators, we use two alternative methods of housing. For short-term studies involving infectious agents, we use class II biosafety cabinets (Figure 2.2). The experimental mice are aseptically removed from the isolator and placed into microisolator cages (i.e., closed cages with a HEPA-filtered top) in a surface-sterilized biosafety cabinet or "hood." The hoods are designed to prevent entry of particulate matter, and with care, they can be maintained in a sterile condition for several weeks. All

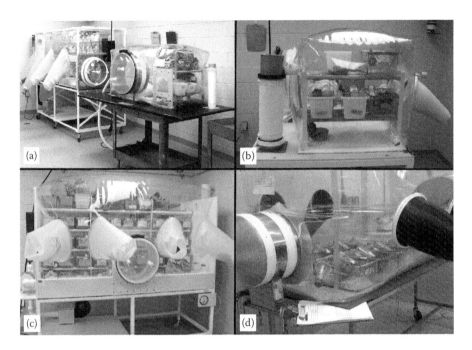

Figure 2.1 Examples of isolator styles (see the Appendix for manufacturer information). (a) Rear: Large isolator with external supports and end port manufactured by Harlan. Front: Small experimental isolator manufactured by CBClean. (b) Side view of small experimental isolator manufactured by CBClean. (c) Large self-supporting isolator with front port manufactured by CBClean. (d) Large isolator with end port and gloves on both sides manufactured by Standard Safety.

animals in hoods are handled with strict surgical asepsis, and we have been able to maintain several different experimental groups in the same hood without cross contamination. The hoods are convenient because the animals are more easily accessible than in the isolators, experimental manipulation is easier, and more than one experimental group can be kept in a single hood. That said, note that the hoods prevent contamination using airflow only, and any interruption of that flow breaks the barrier. For that reason, animals in hoods are not considered truly germ-free, and we do not expect maintenance of sterility beyond a week or two.

The third option for housing germ-free mice is a new technology developed by Tecniplast, called the ISOcage P (Figure 2.3).[3] This caging system was developed for containing biosafety level 3 (BSL-3) and BSL-4 pathogens, and it consists of a completely closed system in which each cage is essentially a mini-isolator. The cages on

Figure 2.2 Class II biosafety cabinet.

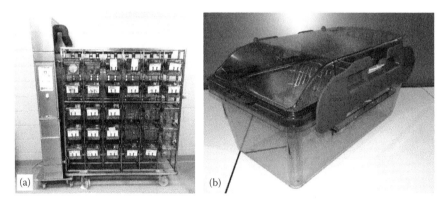

Figure 2.3 (a) Tecniplast ISOcage P rack designed to deliver HEPA-filtered air to cages via a sealed system. (b) Individual ISOcage. The cage is airtight and seals when removed from the rack, essentially acting as a self-contained germ-free isolator.

the rack are airtight and receive double–HEPA-filtered air that flows through the cage and exits via another HEPA filter. When the cage is removed from the rack, it automatically seals, preventing any air movement in or out. We have found these cages useful for experiments with multiple groups inoculated with different microorganisms because there is essentially no chance of cross contamination between cages. Husbandry and experimental procedures, however, must be performed in a class II biosafety cabinet to allow the cage to be opened while minimizing risk of contamination. For this reason, isocages are more prone to accidental contamination than are the Trexler isolators.

Regardless of the housing used, the principal aim of germ-free husbandry and research is asepsis. If an isolator becomes contaminated, it must be completely dismantled, all of the mice culled, and the isolator and its contents taken apart, resterilized, and reassembled. For this reason, prevention of contamination is essential in all aspects of germ-free husbandry. All materials must be sterilized by autoclaving, surface sterilization, irradiation, or gas sterilization, and moving animals and supplies requires repeated surface sterilization throughout the procedure. Biological monitoring must be consistent, and careful records must be kept. Isolators and related equipment must be examined regularly for damage or leaks and kept in repair.

The purpose of this manual is to provide illustrated, step-by-step instructions for the procedures that we use to ensure that our mice remain sterile and healthy. Some of the details described may differ somewhat from procedures in other institutions or in using different equipment. However, the general principles are the same and should be applicable to many, if not all, institutions. These procedures are based on our experience in determining what works in our laboratory. We intend that they be useful for others in setting up their own laboratories or in seeking alternative approaches and procedures.

References

1. Reyniers J, Trexler PC, Ervin RF. Rearing germ-free albino rats. In: Reyniers JA, Ervin RF, Gordon HA, eds. *Lobund Reports*. Volume 1. Notre Dame, IN: University of Notre Dame Press, 1946:1–84.

2. Trexler PC, Reynolds LI. Flexible film apparatus for the rearing and use of germfree animals. *Applied Microbiology* 1957; 5:406–412.

3. Hecht G, Bar-Nathan C, Milite G, Alon I, Moshe Y, Greenfeld L, Dotsenko N, Suez J, Levy M, Thaiss CA, Dafni H, Elinav E, Harmelin A. A simple cage-autonomous method for the maintenance of the barrier status of germ-free mice during experimentation. *Laboratory Animals* 2014; 48:292–297.

Equipment and Terminology

The basic equipment for running a germ-free mouse facility includes the germ-free barrier (usually bubble-type isolators), caging and cage rack supports for use inside the isolators, and equipment for sterilizing supplies (see Chapters 5 and 8). Most supplies are autoclaved in cylinders that are designed to permit aseptic transfer directly into the isolators (see Chapter 8). Surface sterilization is accomplished using a commercial atomizer driven by an air compressor. The atomizer produces a fine mist, essentially vaporizing the sterilant, which is sprayed on surfaces and into isolator spaces. The process of sterilizing the inside of an isolator using vaporized sterilant is referred to as *fogging*, and it accomplishes two things: It ensures that sterilant contacts all the surfaces within the space being fogged, and it inflates the isolator, facilitating detection of leaks (see Chapters 6, 7, and 9).

Figures 3.1–3.8 are examples of major equipment and their uses.

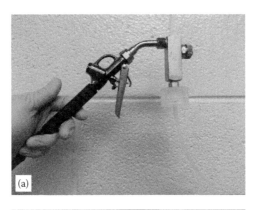

Figure 3.1 (a) Atomizer used for fogging isolators. A sterilant bottle is attached to the screw cap, and the black hose is attached to the air compressor. (b) Atomizer fogging an isolator port.

Figure 3.2 Supply cylinder used for bulk autoclaving of supplies (see Chapter 8).

Figure 3.3 Supply isolator. A supply isolator essentially quarantines organic husbandry supplies. It houses the supply of food and bedding for a breeder isolator as well as a few surveillance mice (Chapter 8). The supply isolators can store up to three cylinders. Following deposit of the cylinder contents, supply isolators are monitored for sterility three times at weekly intervals prior to transfer into a breeding isolator. This practice greatly reduces the risks of contamination of the main colony. Supply isolators never come in contact with another isolator other than the assigned breeding isolator.

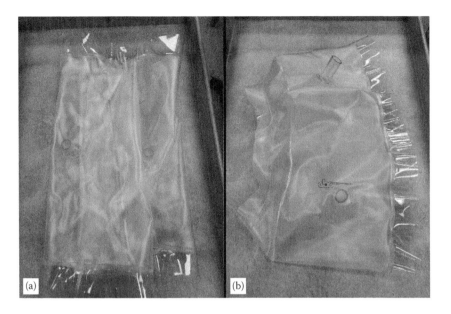

Figure 3.4 Transfer sleeves are used to transfer materials between isolators or from supply cylinders to isolators. We use two types of transfer sleeves: (a) 18"-to-18" sleeves are used to transfer materials between isolators with 18" ports. These sleeves are plastic tubes, 18" in diameter from end to end, and come in various lengths. (b) 18"-to-12" step-down sleeves are 18" at one end and narrow to 12" at the other end. They are used to transfer between isolators with unequal port sizes.

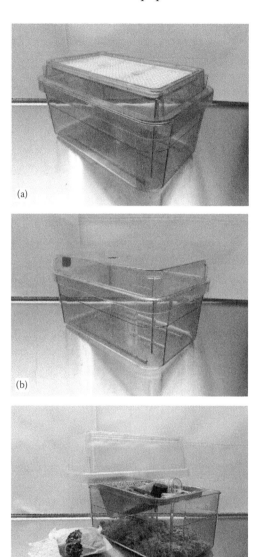

Figure 3.5 Types of caging. (a) Standard microisolator cages with HEPA-filtered lids are used in isolators and in biosafety cabinets. HEPA filters diminish but do not eliminate cross contamination between cages in the same isolator or hood. (b) Solid-top cages are used to transfer mice aseptically between isolators (see Chapter 11). (c) Typical cage kit for use in biosafety cabinets. Cage, food rack, water bottle, and bedding are assembled and autoclaved as a unit for entry into biosafety cabinets (see Chapter 12).

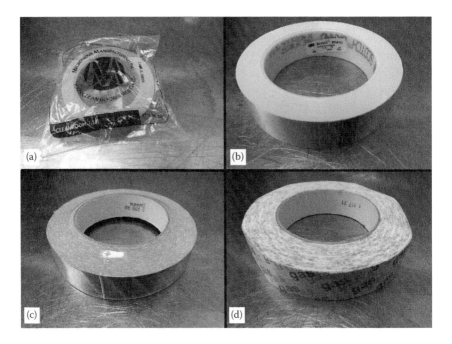

Figure 3.6 Tape. We use four kinds of tape. Nylon tape (a) and high-temperature tape (b) are for wrapping filters and isolator components (see Chapter 6). Autoclave tape (c) and ethylene oxide tape (d) are indicator tapes for steam and gas sterilization, respectively.

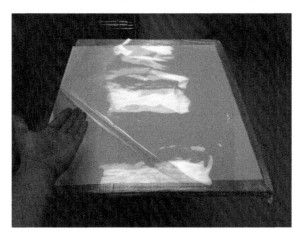

Figure 3.7 Polyester film is used to seal cylinders and filters for autoclaving. We use 22" × 22" precut squares of Mylar® brand film (Dupont Tejjin Films).

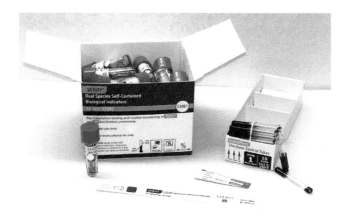

Figure 3.8 Sterilization indicators. Verify® ampoules, steam strips, and Steraffirm® tubes are used to verify the sterilization process (see Chapters 5 and 8).

Personal Protective Equipment (PPE)

Introduction

In the germ-free laboratory, personal protective equipment (PPE) is used either for protection of animals and materials from contamination or for protection of the user from contact with toxins or pathogens. We use three levels of PPE. PPE used for protection of animals in isolators is fairly minimal because the isolator itself prevents contact with microorganisms. For this reason, normal operation requires only latex gloves, plastic gown, and possibly face mask or hair bonnet for preventing contamination of the outside of the isolator in case the isolator itself is damaged (all-purpose PPE). Handling of germ-free animals in class II biosafety cabinets requires a surgical level of sterility to prevent any contact between the user and the cage, animals, surfaces, or anything else (sterile PPE). Finally, respirator PPE, used to protect the user from toxins or infectious agents, consists of full-body protection (gloves, gown, hair bonnet) and a respirator.

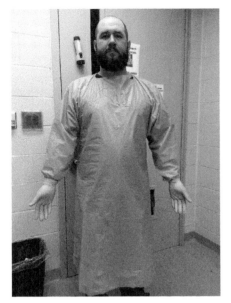

Figure 4.1 All-purpose PPE.

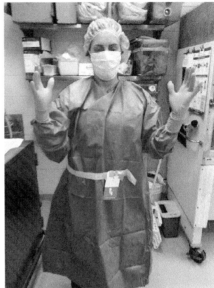

Figure 4.2 Sterile PPE.

Types of PPE

Wearing the correct PPE for the proper use is important for personal safety and for the success of the studies. The three main types of PPE that we use are as follows:

1. All-purpose PPE includes plastic gown, gloves, and optional face mask (required when assisting in aseptic technique). This type of PPE can be worn for isolator operation and cleaning (Figure 4.1).

2. Sterile PPE includes sterile gloves, sterile gown, hair bonnet, and face mask. This PPE is for manipulating germ-free or gnotobiotic mice in a class II biosafety cabinet (see Chapter 12) (Figure 4.2).

3. Respirator PPE includes respirator, eye protection, gloves, and a plastic gown. This PPE is for using volatile sterilant (Alcide®, Clidox®, Spor-Klenz®, or other volatile material) to surface sterilize items by washing or by atomizing (see Chapters 5–9) (Figure 4.3).

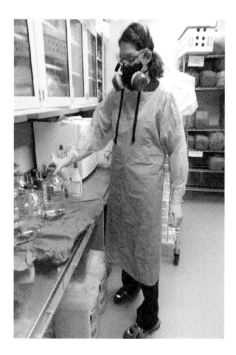

Figure 4.3 Respirator PPE.

Applying Sterile PPE

For the most part, use of PPE is the same in the germ-free facility as in any other laboratory animal facility. Because of the need for surgical asepsis, however, a special technique is needed when applying sterile PPE.

Supplies

- Sterile gloves
- Sterile gown
- Face mask
- Hair bonnet

Procedure

1. Aseptically open the sterile gown and gloves. Make sure the packaging does not obstruct access to the gloves and gown (Figure 4.4).

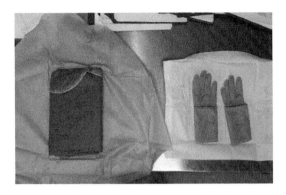

Figure 4.4 Arrangement of supplies for sterile PPE.

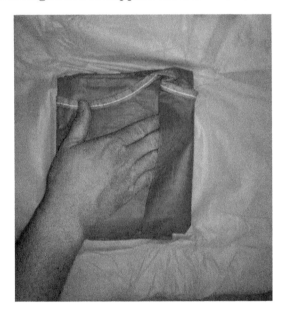

Figure 4.5 Insert dominant hand under the flap leading into the sleeve of the gown.

2. Insert your dominant hand under the flap leading into the sleeve of the gown (Figure 4.5).

3. Once you have inserted your arm in the sleeve, carefully place the other arm in the opposite sleeve. Move your hand into the sleeve, stopping at the cuff. Step back and let the gown unfold over your chest. Be careful not to touch the outside of the gown. If the gown falls or comes in contact with anything, consider it contaminated and start again with a new gown (Figure 4.6).

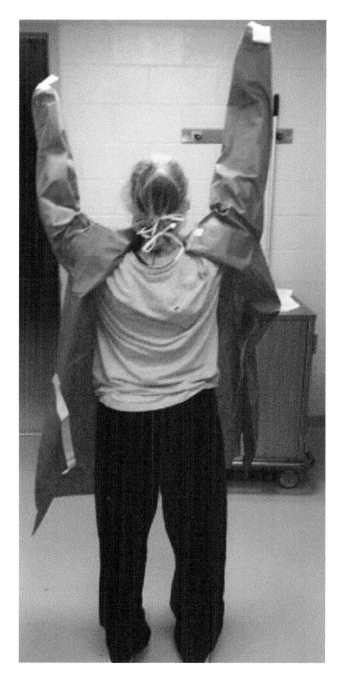

Figure 4.6 Carefully place the other arm in the opposite sleeve. Move your hand into the sleeve, stopping at the cuff. Step back and let the gown unfold over your chest.

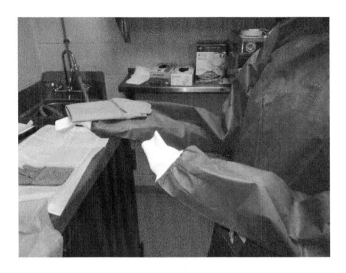

Figure 4.7 From within the sleeve, grasp the dominant glove with your nondominant hand.

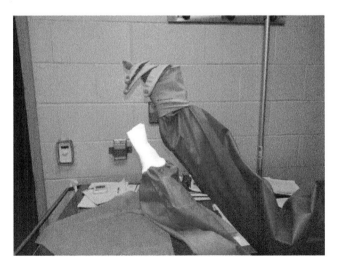

Figure 4.8 Carefully manipulate the sterile glove onto your dominant hand, at the same time manipulating your dominant hand through the cuff and into the glove.

4. From within the sleeve, grasp the dominant glove with your nondominant hand (Figure 4.7).

5. Carefully manipulate the sterile glove onto your dominant hand. At the same time, manipulate your dominant hand through the cuff and into the glove (Figure 4.8).

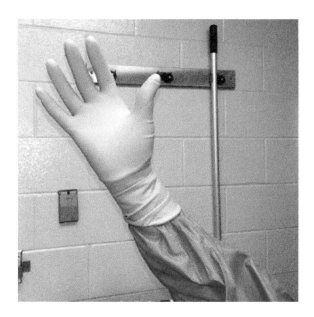

Figure 4.9 You can now freely touch the other sterile glove with your gloved hand.

Figure 4.10 Place the nondominant glove on your nondominant hand.

6. Once you have the sterile glove fitted, you can now freely touch the other sterile glove with your gloved hand (Figure 4.9).

7. Pick up the nondominant glove and place it on your nondominant hand, which is still inside the sleeve of the gown (Figure 4.10).

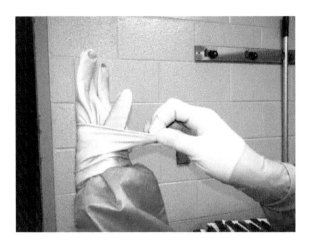

Figure 4.11 Work the glove on your nondominant hand until it is properly fitted.

8. Work the glove on your nondominant hand until it is properly fitted (Figure 4.11).

With full sterile PPE, you are now ready to perform aseptic manipulations in a class II biosafety cabinet.

Sterilants and Sterilization

Introduction

Sterilization is the backbone of the germ-free laboratory. Before they enter the isolators, hoods, or isocages, all materials must be completely free of bacteria, viruses, fungi, and any other microorganisms and must be kept sterile at all times. Any method of sterilization must be effective against not only viable bacteria but also bacterial and fungal spores, nonenveloped viruses, and so on. In short, sterilization must be absolute.

Sterilization Methods

There are four main types of sterilization: cold sterilization (using a liquid chemical sterilant), autoclaving, ethylene oxide (ETO; gas sterilization), and irradiation. The appropriate method of sterilization depends on the type of material (Figure 5.1). Plastics, like polypropylene, Nalgene, and polycarbonate, can withstand high temperatures and can be autoclaved. Nonporous materials can be cold sterilized. Sensitive equipment like electronics can only tolerate ETO sterilization. Organic material like bedding and food must be autoclaved in a sealed metal cylinder. Irradiation may be used for sterilization of food, or irradiated food may be purchased. We find, however, that depending on the source and method of irradiation, irradiated food may not be completely sterile, and we prefer to autoclave food and bedding (see Chapter 8). In this chapter, we describe preparation and sterilization of supplies.

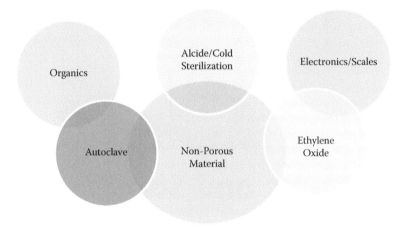

Figure 5.1 Different ways that material can be sterilized. Nonporous material means glass, metal, and polycarbonate caging.

Cold Sterilization

Supplies

- Cold sterilant
- Gloves
- Respirator
- Face mask
- Gown
- Measuring container
- Chemical bin
- Cart

Liquid sterilant is used to sterilize isolator canopies, hoods, and anything entered into an isolator except sensitive equipment and electronics. The sterilant that we use is chlorine dioxide. The brand that we use is Alcide®. It is a 1:4:1 mixture of sodium chlorite, distilled water, and lactic acid, respectively. Other sterilants that can be used for cold sterilization in the germ-free laboratory include Clidox® (a different brand of chlorine dioxide), 10% bleach, peracetic acid (either alone or combined with hydrogen peroxide [Spor-Klenz®], and a new product called Steriplex®. Our choice of chlorine dioxide is based on its wide range and its relatively rapid killing time. Bleach

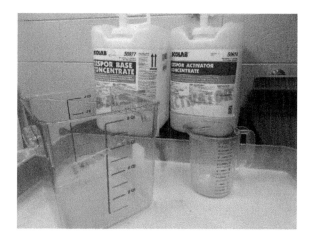

Figure 5.2 Components of chorine dioxide sterilants.

is inexpensive but corrosive, and peracetic acid is no longer used because of its toxicity.

Respirator personal protective equipment (PPE) use is needed for mixing chemicals. The steps for preparing Alcide for use are as follows (use of Clidox is similar):

1. Thoroughly wash and rinse all containers before mixing chemicals (Figure 5.2).

2. Measure the proper quantity of distilled water. For this mixture, we measure 4 cups of distilled water, 1 cup of sodium chlorite (base), and 1 cup of lactic acid (activator). Adjust the volume according to your needs. Be sure to rinse the cup before using it to measure chemicals (Figures 5.3 and 5.4).

3. Alcide can only be used the day that it is prepared.

Preparation and use of other chlorine dioxide sterilants may vary. Follow manufacturers' instructions (see sources, Appendix).

Note: All materials should be autoclaved prior to cold sterilization. After autoclaving, be sure to pay attention to the condition of the materials you plan to enter in the port. Examine the seal "dip" on the water bottles (see water bottle sterilization in Chapter 8). If the dip is not sufficiently concave, do not use the bottle. Pay special attention to the blue wraps of the autoclaved equipment. Make sure there are no tears in the fabric that might have compromised the sterility of the item. If in doubt regarding the sterility of the item, start over.

Figure 5.3 Steps in mixing sterilant are shown in Figures 5.3 and 5.4 (see text).

Figure 5.4 Final mixture.

Procedure

1. Start with a clean cart. Wipe the work area with chlorine dioxide (Figure 5.5a).

2. Soak the clean, autoclaved terry towels in the chlorine dioxide bath. Wipe the outside of the container with chlorine dioxide (Figure 5.5b).

3. Wash and soak the desired materials (sleeves, water bottles, caging materials) for 60 minutes. Figure 5.5c demonstrates

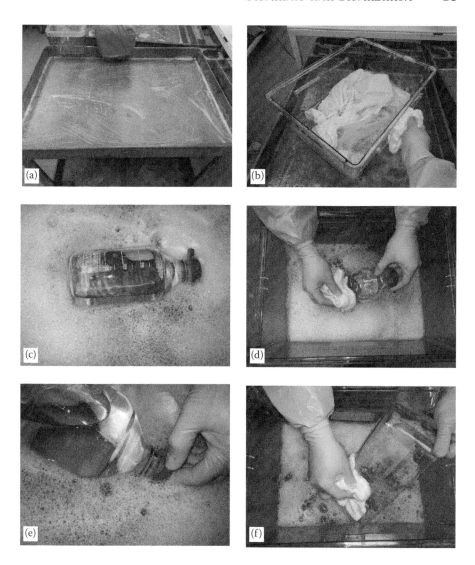

Figure 5.5 Steps for cold sterilization (see text).

soaking of a presterilized water bottle prior to entry into an isolator.

4. Water bottles receive special attention. First, wash the glass (Figure 5.5d). Then, ensure that the rubber lid that seals the flask receives sufficient sterilant contact. To do this, submerge the top of the water bottle and lift the edges of the seal slightly (Figure 5.5e). Be extremely careful not to break the seal. Make sure the entire surface area is exposed to the

sterilant. Then, scrub the top lid. Let the bottles soak as you prepare the port (see Chapter 7).

5. Caging material is always autoclaved before cold sterilization. Examine the blue wrap for damage prior to removing the caging material for cold sterilization. Scrub the caging (or any other nonporous material) and place it to the side of your workspace (Figure 5.5f). Cages and accessories should be cold sterilized as soon as possible before they are placed in the port for entry into the isolator.

Wrapping Materials for Autoclaving

Supplies

- Assembled cage kits or other supplies to be autoclaved. (Note: Cage assembly may vary depending on use. We assemble standard, complete caging kits for use with gnotobiotic mice in biosafety cabinets; see Chapter 3.) Husbandry equipment that is designated for new isolators can be autoclaved without wrapping.
- Kimberly© wraps or similar surgical wraps
- Autoclave tape
- Steam strip indicator

Procedure

1. Position the blue wrap on a smooth, flat surface in a diamond shape. Place the item to be wrapped in the center (Figure 5.6a).
2. Grasp the south point and cover the center of the kit (Figure 5.6b).
3. Hold the south point with your elbow and fold the west and east over the center (Figure 5.6c).
4. Turn the box so that the north side is closest to you (Figure 5.6d).
5. Tightly fold the north side into the fold that you made with the east and west points (Figure 5.6e).
6. Insert a steam strip indicator under the flap. Tape the fold with autoclave tape (Figure 5.6f).

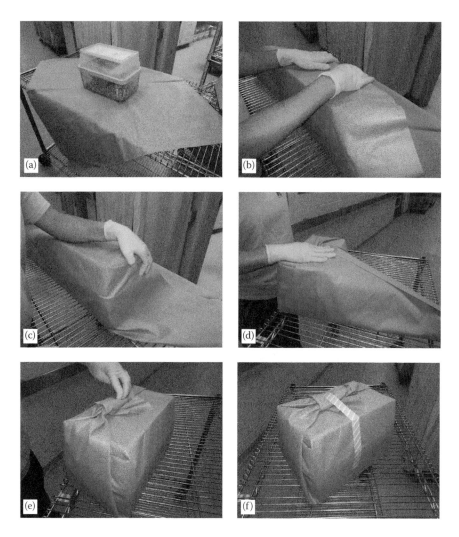

Figure 5.6 Steps for wrapping materials prior to autoclaving (see text).

7. If the cage is to be used in a biosafety cabinet (see Chapter 12), repeat the process, covering the first wrap with a second wrap. Items to be entered into isolators after setup or into biosafety cabinets (see Chapter 12) are double wrapped to permit aseptic removal later. Details of the autoclaving procedure are in Chapter 8.

8. The same wrapping technique is used to prepare items to be gas sterilized. Instead of the autoclave tape and steam strips, use an ETO indicator strip and gas sterile tape.

Figure 5.7 Ethylene oxide sterilizer.

Gas Sterilization (Ethylene Oxide)

Most major hospitals have an ETO sterilizer (Figure 5.7). ETO is used to sterilize items sensitive to heat or chemicals. (Items like electronics, plastics, and sensitive equipment must be sterilized using an ETO sterilizer.) It is also used to sterilize instruments to prevent degradation/dulling. In this section, we discuss how to prepare items for ETO sterilization, operation of the machine, and safety.

Supplies

- ETO sterilizer
- ETO canister
- Sterilization pouch (self-sealing)
- ETO verification strips
- ETO tape
- Kimberly© blue wraps

Procedure

1. Small items for ETO sterilization are placed in peel pouches that can be easily opened without compromising the sterility of the contents (Figure 5.8). Choose the correct size peel

Figure 5.8 Peel pouches with indicator strips. The top pouch contains sterilized tubes and a processed indicator strip. The bottom pouch contains an unprocessed strip.

Figure 5.9 Entering item to be sterilized and indicator strip into the peel pouch.

    pouch for the item. In Figure 5.8, the small pouch contains sterilized sample tubes. Notice the color difference between the processed indication strip (small package) and the unprocessed strip below it.

2. Insert the desired item and the ETO verification strip into the peel pouch (Figure 5.9). Be sure that the item is not so large in diameter that the peel pouch will not seal correctly.

3. Seal the pouch by peeling the cover strip off the adhesive edge and pressing closed (Figure 5.10). Carefully seal the end. Hold the end up to a light source to reveal any creases or folds, which may result in a bad seal. A bad seal may compromise the sterility of the contents.

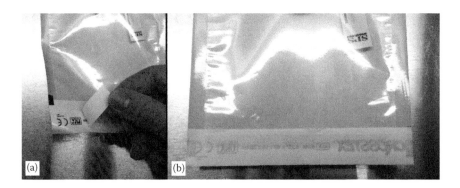

Figure 5.10 (a) Seal the peel pouch by removing the adhesive strip and pressing closed. (b) Correctly sealed pouch with no folds or creases, ready to be sterilized.

Figure 5.11 Placement of items in the sterilizer.

4. Inspect the ETO sterilizer to make sure there are no other supplies left by the previous user. Then, place your items such that there are no items in direct contact with each other (Figure 5.11). Remember that ETO gas will need to penetrate the items, and spacing them is crucial.

5. Check to make sure the used canister of ETO is removed (Figure 5.12a).

6. Retrieve a new, full ETO canister from the storage box (Figure 5.12b). *Use care:* **Ethylene oxide is extremely flammable!**

7. Place the new canister firmly in place (Figure 5.12c).

Figure 5.12 Replacing the ETO canister. (a) Ensure that the used canister is removed. (b) Canisters should be stored in a flameproof cabinet. (c) Replace the canister.

8. Usually there are two types of cycles:

    a. Low, 100°F (40.5 hours running time)

    b. High, 130°F (13 hours running time)

Select the cycle based on the sensitivity of the material. High temperature is used for instruments, and low temperature is used for electronics.

6

Isolator Setup

Introduction

As noted previously, we use soft-sided, bubble-type Trexler isolators. We do not have experience with solid-sided isolators, and those are not addressed here. Trexler isolators can be made to order in many different sizes and shapes, but all contain the same basic components: plastic canopy, entry port, gloves, and inflow and outflow filtered air supply (Figure 6.1). The polyurethane canopy is inflated by a ventilation motor that passes air through a HEPA filter. The canopy inflates to a positive pressure, then exhausts through a second HEPA filter. The only access into the isolator is through the port. The port has a double air-lock system consisting of two barriers: the inside cap and the outside cap. When the isolator is in use, only one of these barriers is removed at a time.

The isolators we use (and recommend) have HEPA filters on both inflow and exhaust pipes. HEPA-filtered exhaust protects workers from any pathogens that may be used experimentally in the isolators and prevents contamination of the isolator in case of pressure loss and reversed flow of air into the isolator. The type of filter may vary between labs and between isolator types, but in all cases filters must be assembled, wrapped, and sterilized before use. Once sterile, they can be used for replacement (done yearly for large breeding isolators) or to establish new isolators.

We use two styles of bubble isolators. **Small isolators** (5.5' × 3') have a single port and a single pair of gloves and can be used by one person at a time (see Figure 6.1). We use them mostly for short-term

Figure 6.1 Typical Trexler-type isolator with the major parts labeled.

Figure 6.2 Large breeding isolator used in our laboratory.

experiments and for storing supplies. Ours can comfortably house up to 10 cages and supplies. Similar styles can be custom made in either smaller or larger sizes. Some laboratories use individual smaller isolators that can be stacked. This is useful for conducting long-term experiments with many small groups of mice. Filters and gloves cannot be changed on small isolators when they are in use; the isolators must be taken down and resterilized when replacements are needed. **Large isolators** (breeding isolators) have two pairs of gloves, allowing two people to work in the isolator at the same time

(Figure 6.2). They can house up to 50 cages, and we use them primarily to house our breeding colony. Other sizes and styles of isolators are available (see Figure 2.1). Using proper technique, filters and gloves can be replaced on active large isolators. The basic construction of the different styles is similar, but details vary depending on the complexity of the airflow delivery, number of ports and gloves, and so on.

Isolator assembly requires care and attention to detail. Component parts must be fitted precisely and care must be taken not to damage the canopy or any of the parts. The isolators described in this chapter are commonly used styles manufactured by CBClean (see Appendix). The assembly process may be modified to fit alternative isolator styles (see Appendix for sources of isolators), but we include sufficient detail here to indicate the appropriate order of the steps as well as possible trouble spots. This chapter is divided into two parts, describing setup of the smaller isolators in part 1 and the larger breeding isolators in part 2.

Part 1: Small (5.5 × 3 feet) Isolator Assembly

Preparation of Isolator Components

Setting up a new smaller isolator (Figure 6.3) requires considerable preparation time. The first steps involve preparing materials for

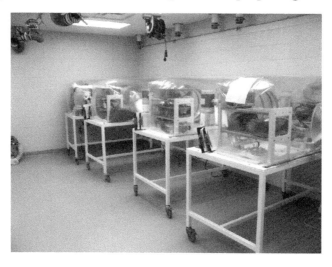

Figure 6.3 Small isolators used for experiments and storage of supplies.

entry into the isolator. Tasks involved include installing HEPA filters, wrapping and sterilizing the exhaust and intake filters and caging equipment, and washing and drying plastic components. Strict aseptic technique is not needed when you install the components of the isolator, but all components should be at least clean if not sterile prior to assembly. Caging material should be preautoclaved. The caging components can be placed on an autoclavable cart, covered with a linen cloth, and autoclaved in bulk. Be sure to inspect every cage component for damage. Look for sharp, broken edges on glass bottles and broken welds on wires. Cold sterilization (chlorine dioxide) is essential only when attaching certain components (the filter ports, outside caps, gloves, and glove cuffs) to the isolator canopy. Each of these components connects with portions of the canopy that will be sealed off for the entire duration of the isolator's use and therefore must be sterile during setup. The internal components will be cold sterilized later (see discussion that follows).

Supplies
- Regular personal protective equipment (PPE) (Respirator PPE should be used when chlorine dioxide is in use.)
- Sterilized caging
- Cold sterilant (approximately 3–4 gallons)
- Sterile terrycloth towels
- Atomizer
- Air compressor (minimum of 80 psi)
- Clean cart/table space
- Nylon tape
- Chemical respirator
- Autoclaved tools:
  - 5/16 socket screwdriver
  - Phillips screwdriver
  - Allen wrench
- Isopropyl alcohol
- Hearing protection
- Crucible pliers
- Sterile pen
- Filled, 2 liter water bottle (sterile)

Isolator Components

- Sterile intake and exhaust filters (prepared as described in the next section)
- Two 3" clamps
- One pair of number 9 butyl rubber dry box gloves
- Two 8" plastic rings
- Two 8½" ring gaskets
- Two 20" stainless steel clamps
- Outside cap (with nipples)
- Two inside caps (without nipples)
- Six 18" rubber bands
- Port (18")
- Port clamp (62")
- Canopy port clamp (62")
- Canopy port gasket (smooth)
- Outside cap grooved gasket with outside clamp
- 1" rubber stoppers
- Canopy
- Autoclaved caging (minimum of 15 minutes at 121°C)
- Floor mat
- Cage racks
- Canopy support rod

Preparing Hand-Wrapped Filters

Supplies

- Filter frames
- DW-4 HEPA filtration media
- T square and ruler
- Sharp razor blade
- Autoclave tape
- 22" clamps
- Measuring tape
- Nylon tape
- 5/16 socket wrench

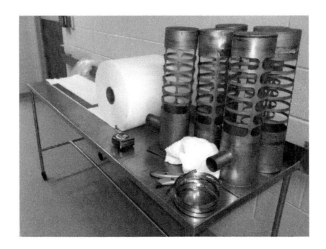

Figure 6.4 Filter frames and supplies for assembling small isolator filters.

Filters for small isolators are hand wrapped in DW-4 HEPA material. Three to four layers of DW-4 are used for intake filters. Two to three layers are appropriate for exhaust filters.

The following are the steps:

1. Start with clean filter frames (Figure 6.4) that have been washed and all adherent material removed. Use a clean and uncluttered workspace.

2. Measure the HEPA filter material (Figure 6.5). For our exhaust filters, we use 36" × 12" lengths, and for the intake filters, we use 52" × 12" lengths. Use a square to cut and a ruler to measure. Make sure the razor blade is new.

3. Straighten the HEPA material and place it into position. The filter material should overlap the edges of the filter ventilation slots by at least ½" (Figure 6.6). Use a strip of autoclave tape to affix the material tightly against the filter and wrap the filter tightly around the frame.

4. Once the HEPA filter is in place, secure it using the 22" stainless steel clamp over the overlap area. Cover the clamp and the edge with nylon tape (Figure 6.7). Do the same for the other edge.

5. Place a piece of polyester film (Mylar®) over the opening of the filter (Figure 6.8).

6. Use an elastic band (we use a hair tie) to secure the film tightly around the opening of the filter (Figure 6.9).

Figure 6.5 Measure and cut HEPA media for filters.

Figure 6.6 Align filter media with filter frame.

7. Wrap the nylon tape once over the edge of the filter opening (Figure 6.10).

8. Cut off the excess polyester film (Figure 6.11) and remove the hair tie.

9. Cover the edge of the nylon tape and wrap the nylon tape toward the filter, then back to the front edge (Figure 6.12). The nylon tape layer should be thinner toward the filter.

10. Wrap the filter in a blue wrap (Figure 6.13; see Chapter 5). Autoclave between 122°C and 127°C for 25 minutes (Figure 6.14).

Figure 6.7 Cover the clamp with nylon tape.

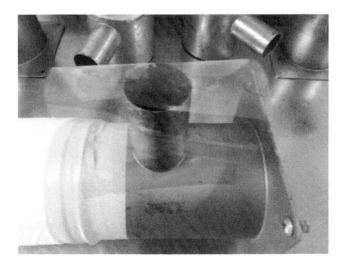

Figure 6.8 Place polyester film over the filter opening.

Figure 6.9 Secure with an elastic band.

Figure 6.10 Tape the film around the filter housing.

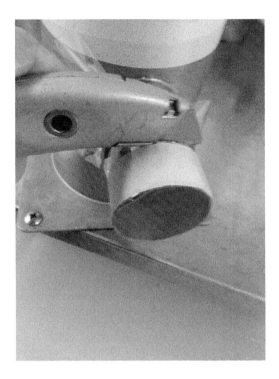

Figure 6.11 Cut the tape and remove excess film.

Figure 6.12 Finish taping the film to the housing.

Figure 6.13 Wrap the filter for autoclaving.

Figure 6.14 Wrapped and autoclaved filters ready for installation.

Installing the Isolator Components

1. Begin by installing the motor, port clamp/mount, and air pipes to the isolator stand or table (Figure 6.15). This is also a good time to clean the stand and inspect the motor.

2. Install the 18" port. Make sure the port clamp is hand tightened (Figure 6.16). The port should be positioned so that about a third of the length is within the interior of the isolator.

Note: **Proper placement is extremely important.** If the port is misaligned, there might not be enough space to properly place the inside cap.

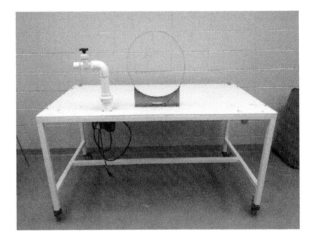

Figure 6.15 Port mount, motor, and air pipes installed.

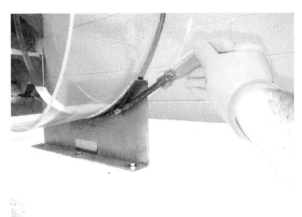

Figure 6.16 Install the port and tighten the clamp.

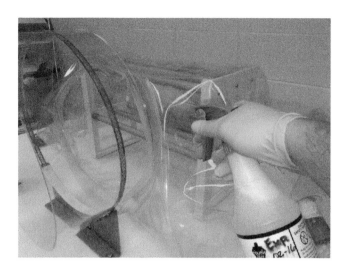

Figure 6.17 Canopy in place with rack inside.

3. Place the canopy and attach it to the port:

    a. Position the clean canopy on the isolator table, assemble the cage rack(s) outside the isolator, and carefully place the racks inside the canopy (Figure 6.17). Keep the rack hardware loose. You will need to cold sterilize the space between the support boards and rack poles prior to final assembly.

    b. Place the autoclaved correct size Allen wrench (or screwdriver) inside the isolator.

    c. Place the smooth 18″ canopy/port gasket on the interior side of the port (the side facing the canopy). Starting at the bottom of the port opening, slowly apply the port opening to the port. The port opening and port are the same dimensions, so you will need to stretch the polypropylene just slightly to fit it over the port. The use of alcohol makes this process a little easier. It will lubricate the polypropylene and disinfect the area (Figure 6.18).

    d. The edge of the canopy port opening should extend to the edge of the 18″ port-mount clamp (Figure 6.19).

    e. Slide the gasket over the polypropylene port opening edge, secure (hand tighten) it with an 18″ stainless steel clamp (the same size as the port mount clamp; red arrow). Use nylon tape to cover the stainless steel clamp (Figure 6.20).

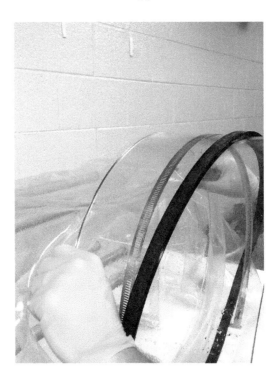

Figure 6.18 Fit the porthole of the canopy over the port.

Figure 6.19 Canopy porthole should extend up to the port mount (green arrow). In this photograph, the inside of the isolator is to the right.

Figure 6.20 Assembled canopy and port. Cover the clamp (red arrow) with nylon tape.

4. Install the number 9 dry box–style, butyl rubber gloves (Figure 6.21). Sterilant is used on all surfaces of each component during installations. For this step you need:

Two 8" plastic rings

Two 20" stainless steel clamps

Two 8" rubber gaskets, 1/2" wide

One 5/16 socket wrench

a. Wash the rings and the canopy material around the glove opening with sterilant. Place the 8" plastic rings on the edge of the 8" openings on the canopy (Figure 6.22).

b. Overlap the canopy material over the 8" rings far enough to cover the outside edge of the rings (Figure 6.23).

c. Wash the entire glove in sterilant.

d. Insert the left glove into the left 8" glove opening of the canopy (Figure 6.24).

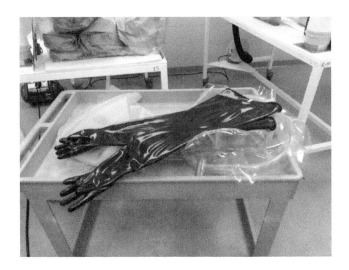

Figure 6.21 Dry box–style gloves ready to be installed.

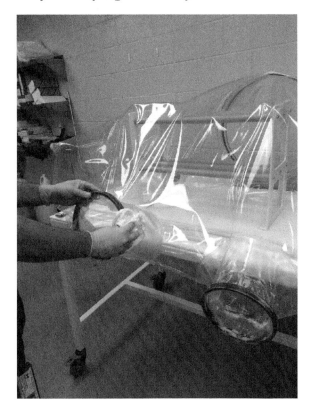

Figure 6.22 After the glove rings are installed, wash the surrounding canopy in preparation for glove installation.

Figure 6.23 Glove hole with glove rings inserted.

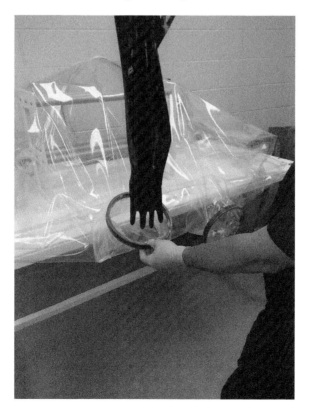

Figure 6.24 Glove insertion.

Figure 6.25 Rubber glove gaskets.

    e. Make sure to position the glove *thumb up*. This positions the glove correctly for use inside the isolator. Overlap the edge of the glove the same way the canopy material overlapped the plastic ring.

    f. Take the 8″ rubber gaskets (Figure 6.25) and slide them over the butyl rubber gloves. The gaskets should fit snugly in the groove of the rings (Figure 6.26).

    g. Make sure that all components are always washed and covered in sterilant.

    h. Carefully place the 8″ stainless steel clamps around the rubber gasket. Hand tighten the clamps (Figure 6.27).

  5. Place the protective mat inside the isolator and lay it out evenly on the bottom.

  6. Place the two inside caps, the outside cap, a sterile pen, water bottle (sterile/sealed), preautoclaved caging, terrycloth towel, and an extra cage filled with approximately 2 liters of sterilant

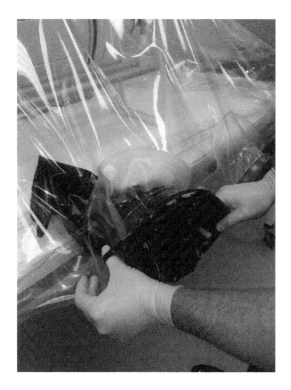

Figure 6.26 Fit the gaskets over the glove cuffs.

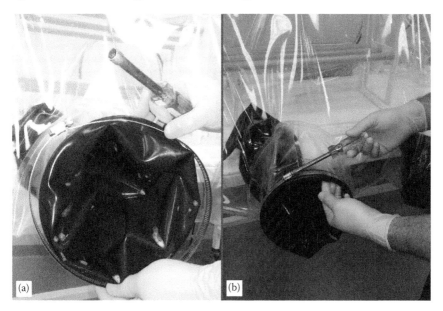

Figure 6.27 Glove clamps. (a) Place the clamp. (b) Tighten the clamp.

Figure 6.28 Isolator with gloves installed and two inside caps placed inside (one extra for emergencies).

in the isolator. **Note:** Two inside caps are placed in the isolator in case of damage to the first cap with attendant loss of pressure (Figure 6.28).

7. Install the filters:

a. Start with presterilized filters (Figure 6.29).

b. Prepare the filter opening on the canopy by washing the area with sterilant (Figure 6.30). Wash the filter seal and surrounding area with sterilant (Figure 6.31).

c. Carefully insert the filter into the filter port (Figure 6.32). Do not break the seal.

d. Cover the base of the filter port with nylon tape before tightening the clamp (Figure 6.33). This will help pad the clamp and keep the polypropylene canopy safe from damage.

e. Hand tighten the 3" stainless steel clamp (Figure 6.34). Cover the clamp screw head with additional nylon tape. This will shield the metal edges.

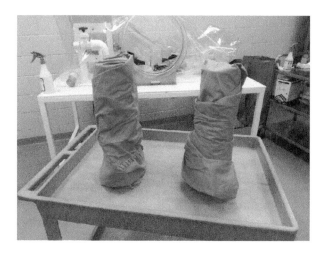

Figure 6.29 Sterilized filters still in their blue wrap.

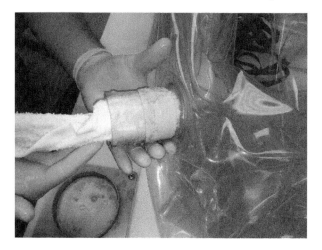

Figure 6.30 Wash the filter port.

    f. Repeat the same steps for the intake filter.

8. Install the canopy rod (Figure 6.35).

Inflating and Sterilizing the Isolator

1. Make sure you have the following clean supplies inside the isolator:

- Preautoclaved caging equipment (with number of cage setups depending on the size of the isolator and its intended use)

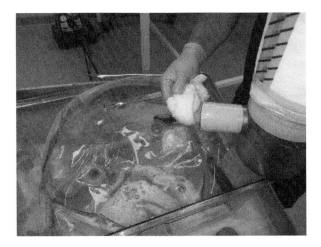

Figure 6.31 Wash the filter seal.

Figure 6.32 Insert the filter into the port.

- Six rubber bands (ethylene oxide [ETO] sterilized)
- Sterile water
- Mat
- Crucible pliers (preautoclaved)
- Sterile pen
- Two inside caps
- Extra cage filled with approximately 2 liters of sterilant
- Large, sterile terrycloth towel
- Allen wrench (preautoclaved)

Figure 6.33 Cover the canopy filter hole with nylon tape before placing the clamp.

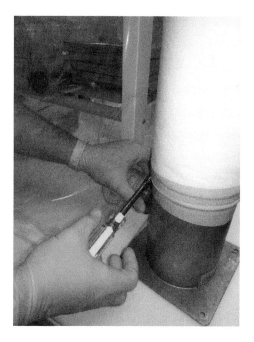

Figure 6.34 Tighten the clamp.

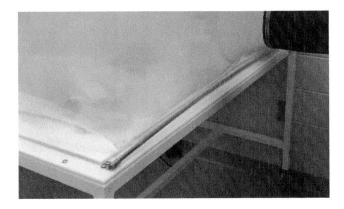

Figure 6.35 The canopy rod at the base of the canopy helps keep the isolator flat.

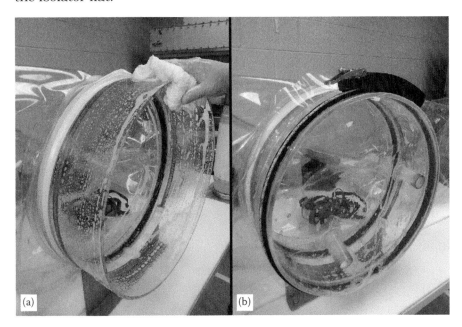

Figure 6.36 (a) Prepare the port. (b) Install the outside cap but not the inside cap.

2. Prepare the port (see Chapter 7) and install the outside cap but do not install the inside cap (Figure 6.36). The isolator is now sealed and ready to inflate.

3. Set the compressor/atomizer to 80 psi.

4. Using the atomizer, inflate the isolator through one of the outside cap nipples (Figure 6.37).

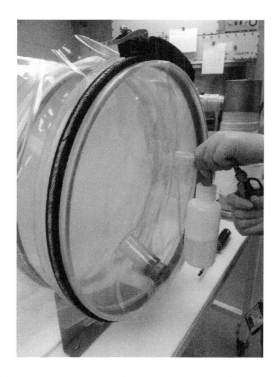

Figure 6.37 Use the atomizer to inflate the isolator with vaporized sterilant.

5. When the gloves are exactly horizontal, plug the outside cap nipples (Figure 6.38).

6. You can make some minor isolator component adjustments, glove inspection, and canopy placement while the isolator is inflating. Most of these adjustments happen spontaneously if you loosen the port, glove, and filter clamps on the isolator filters and port clamps during inflation.

7. After the isolator is inflated, mark the position of the gloves either with a piece of tape or with a photograph.

8. Perform a leak test as follows: Wait at least an hour and recheck the isolator pressure as indicated by any change in the position of the gloves (Figure 6.39). A slight decrease in pressure is completely normal, but if the isolator has a noticeable difference in pressure (drooping of the gloves), retighten the clamps on the isolator components and inflate the isolator again. Wait another hour. If the pressure is noticeably lower again, there is a leak or hole in the canopy or one of the isolator components.

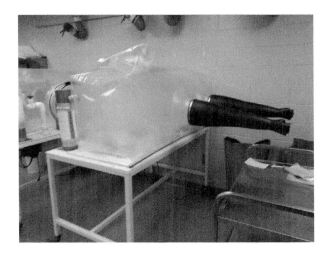

Figure 6.38 Fully inflated isolator filled with vaporized sterilant. Note horizontal position of gloves.

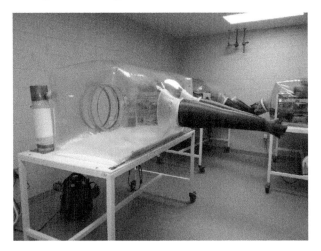

Figure 6.39 Leak test. At the end of an hour, the gloves should remain at or close to their original horizonal position.

9. Once the isolator has passed the leak test, it is safe to continue.

Decontaminate the Inside of the Isolator

1. Deflate the isolator slightly by aseptically removing one rubber stopper on the outside cap (Figure 6.40).
2. Deflate the isolator enough so that you can easily access the inside but not so much that it will collapse (Figure 6.41).

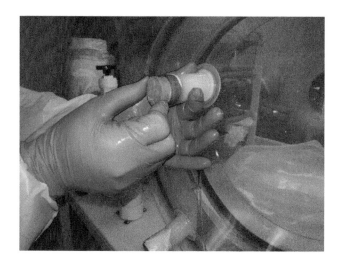

Figure 6.40 Deflate the isolator by removing one of the plugs on the outside cap.

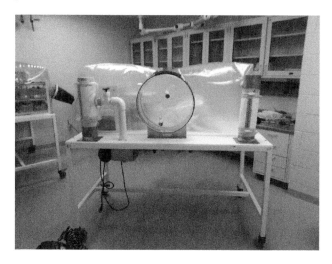

Figure 6.41 Slightly deflated isolator ready for decontamination of the interior.

For the following internal decontamination steps, work inside the isolator via the gloves:

3. Begin by moving all objects to one side of the isolator. Using the sterilant that was placed in the isolator in step 6 of component entry, wash the empty side of the canopy surface. *Pay special attention to the filter ports, corners, and port edges.*

4. Scrub the boxes, water bottles, and wires and carefully place them to the empty/washed side of the isolator. *Be extra careful when washing these components.* There may be sharp metal present that could damage the isolator gloves.

5. Wash the cage rack. Pay close attention to the fasteners. Once the small spaces in between the screws, rods, and side boards of the rack have been exposed to sterilant, tighten the screws with the Allen wrench.

6. Completely submerge smaller components (sipper tubes, 8-ounce water bottles, cage card holders, rubber bands) in the cage containing sterilant. Let them soak for at least an hour. Perform a final inspection on each individual caging component.

7. Repeat the washing on the other side of the canopy.

8. Wash the filter ports, port, and corner seams.

Overnight Leak Test

1. Fully inflate the isolator again and seal the outside cap with rubber stoppers. Use nylon tape to secure the rubber stoppers. This step is similar to the first leak test described previously. Leave the pressurized isolator overnight.

2. After 24 hours, examine the pressurized isolator. A slight deflation is expected, but there should be minimal change in the position of the gloves.

3. As done previously, aseptically remove one rubber stopper from the outer cap until the isolator is slightly depressurized, then replace the stopper (see Figures 6.40 and 6.41). This will help manipulate objects within the isolator. The partly deflated isolator canopy should remain sufficiently pressurized to support itself.

4. Attach the intake filter bag around the intake HEPA filter (Figure 6.42). Make sure the bag is sealed with nylon tape.

Start the Isolator

1. From inside the isolator, place the inside cap on the port (Figure 6.43). Use a terrycloth towel as a grip because the cap will be extremely slippery. Secure the cap with a rubber band.

2. Plug in the ventilation motor.

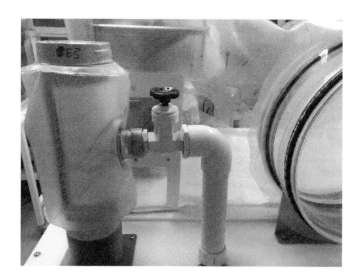

Figure 6.42 Filter bag placed around intake filter.

Figure 6.43 Install the inside cap.

3. Open the air valve (Figure 6.44). Check to make sure the isolator has enough positive pressure to support the canopy.

4. Use a sterile pen or the edge of a cage card holder (any sharp object) to carefully tear through the film seal of each filter (Figure 6.45).

Note: Excess chlorine gas will exit the isolator at this time. Be cautious and make sure there is proper ventilation in the room.

Figure 6.44 Open the airflow valve and ensure that the isolator is under positive pressure.

Figure 6.45 Tear through the polyester film seal on the filters.

5. Use a towel to wipe up any excess sterilant in the isolator and ring out the excess sterilant into an empty cage.

6. Wait 24 hours.

7. The next day, use the sterile water entered during component entry, step 6, to rinse off the interior of the isolator.

8. Put the excess sterilant in the empty water bottle and seal the bottle for later removal.

The isolator is now ready for an initial supply cylinder deposit (see Chapter 9).

Part 2: Large Isolator Assembly

Introduction

Soft-sided bubble isolators are not manufactured; they are all hand-made. Because they are handcrafted, they are all slightly different. This section is specific to the large breeding isolators (Figure 6.46) that we order from CBClean (see sources in the Appendix). We find the large CBClean isolators are ideal for long-term housing of our breeding colonies. The breeding isolators are similar to the smaller isolators but differ in the filter assembly and the type of gloves used. Thus, the setup is similar to the setup and sterilization of the small isolator described in part 1 of this chapter but is a bit more complex.

The suggestions that follow are based on our experience with these particular isolators. They may be modified according to specific isolator features.

As described previously, preparation (autoclaving supplies) of isolator components is necessary.

Supplies
- Regular PPE
- Sterilant (approximately 3–4 gallons)
- Sterile terrycloth towels
- Atomizer
- Air compressor (80 psi)
- Clean cart/table space
- Nylon tape

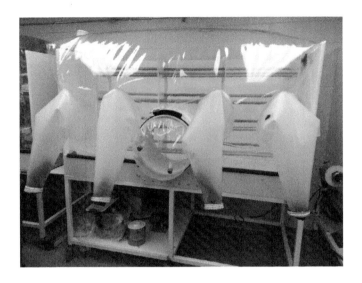

Figure 6.46 Large breeding isolator made by CBClean.

- Chemical respirator
- Autoclaved tools:
  - 5/16 socket screwdriver
  - Phillips screwdriver
  - Wrench
- Isopropyl alcohol
- ETO-sterilized exhaust and intake HEPA filters
- Exterior polypropylene piping components
- Number 9 butyl gloves
- Polytetrafluoroethylene (PTFE) plumber's tape
- Intake housing
- Two interior plugs
- Internal cage rack system
- Outside cap (with nipples)
- Two inside caps (without nipples)
- Six 18" rubber bands
- Port (18")
- Port clamp (62")
- Canopy port clamp (62")

- Canopy port gasket (smooth)
- Outside cap grooved gasket with outside clamp
- 1" rubber stoppers
- Canopy
- Autoclaved caging (minimum of 15 minutes at 121°C)
- Floor mat
- Cage racks
- Two sterile pens
- Sterile water bottle
- Four plastic, grooved glove cuffs
- Twelve 6" O-rings

Preparation

There are three important sterilization steps to prepare for the aseptic assembly of the large isolator:

1. Sterilize the exhaust and intake filters.
2. Sterilize the air supply piping system.
3. Prepare the supplies to be placed into the isolator.

These preparation steps can take several days to accomplish.

Sterilize the Filters

The filters used for the large isolators are manufactured (CBClean) (Figure 6.47). They contain glues and plastic and cannot be autoclaved. Use ETO gas to sterilize the filters. The intake filter is about half the size of the exhaust filter.

1. Cover the filter threads with low-density PTFE plumber's tape prior to sterilization. This will create a seal when the filter is installed (Figure 6.48).
2. Once the threads are wrapped, wrap the filter for sterilization:
   a. Place the filter and an ETO indicator in the center of a 24" × 24" blue wrap (Figure 6.49).
   b. Fold the bottom corner of the blue wrap over the filter (Figure 6.50).
   c. Fold the right corner of the blue wrap to the center (Figure 6.51).

Figure 6.47 Intake filter frame for large isolators.

Figure 6.48 Wrap the filter threads with plumber's tape.

    d. Fold the left corner to the center (Figure 6.52).

    e. Turn the filter around and fold the last, top corner to the center (Figure 6.53).

    f. Use ETO tape to seal the wrap (Figure 6.54).

    g. Apply a second wrap using the same technique.

    h. The prepared filters will be processed in an ETO sterilizer for 12 hours at 130°F.

    i. ETO tape changes color from brown (unprocessed) to green (processed) (Figure 6.55).

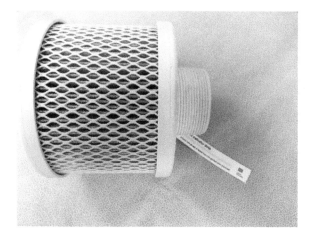

Figure 6.49 Place the filter frame and ETO sterilization strip on a blue wrap.

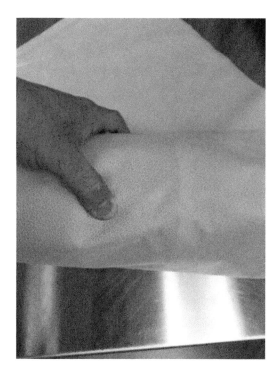

Figure 6.50 Wrapping filters for ETO sterilization; Step 1: Fold the bottom corner of the blue wrap over the filter.

Figure 6.51 Wrapping filters for ETO sterilization; Step 2: Fold the right corner of the blue wrap to the center.

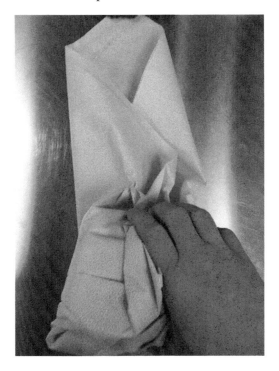

Figure 6.52 Wrapping filters for ETO sterilization; Step 3: Fold the left corner to the center.

Figure 6.53 Wrapping filters for ETO sterilization; Step 4: Turn the filter around and fold the last, top corner to the center.

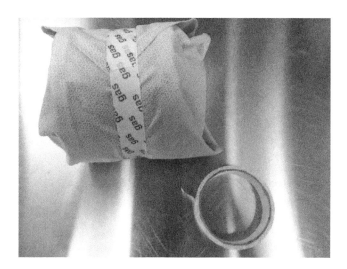

Figure 6.54 Wrapping filters for ETO sterilization; Step 5: Use ETO tape to seal the wrap.

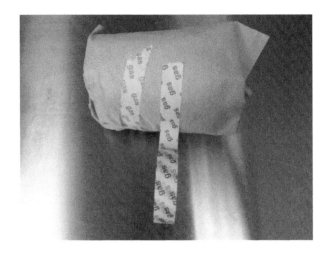

Figure 6.55 Unprocessed (brown) and processed (green) ETO indicator tape.

Sterilize the Air Supply Piping System

Unlike the smaller isolators described previously, the large breeding isolators have external supports and piping for the air intake and exhaust systems. Our air piping system is adapted to fit different diameter filter ports. Because our connections are customized, we classify the piping connections into two components: the 90° assembly and the filter adapter. Sealing the filter ports requires the use of handcrafted polypropylene or rubber gaskets. The pipes used for the air intake and exhaust system are made of polypropylene rather than polyvinylchloride (PVC). Polypropylene pipes are used in a wide variety of agricultural and industrial applications. Unlike PVC, they can withstand harsh chemicals, stress cracking, and high temperatures, all of which are applied to the ventilation system for the large isolators. In contrast, even under no pressure, PVC pipes begin to crack after 3–4 years. This makes the polypropylene far superior than regular PVC for use in isolators.

To prepare and sterilize the piping, apply low-density PTFE plumber's tape on all of the threads of the connections. Inspect every piece of equipment for damage/sharp edges. Place the pieces in a polycarbonate cage with a solid top (Figure 6.56) or blue wrap them and autoclave at 121°C for 35 minutes using the prevac cycle.

Prepare the Supplies

It is best to make sure that all supplies that will be needed are put into the isolator at this point because introducing equipment into

Figure 6.56 Filter piping packed for autoclave.

the isolator after setup is much more labor intensive than entering equipment prior to completion of the setup. The number of cage assemblies to be prepared will vary depending on the isolator size and use. It is important to plan ahead and make a list of the supplies to be included. Place the caging equipment and supplies on an autoclavable cart and cover it with a linen cloth. There is no need to individually wrap the caging equipment for breeding isolator setups (Figure 6.57).

Isolator Assembly

Begin with a clean isolator setup as described in part 1. The canopy has been centered and attached to the port (see the discussion of installing the isolator components in part 1, step 3). An internal cage rack has been assembled inside the isolator. Keep the rack hardware loose. At this point, the port is open, as are all of the other openings for installation of components (Figure 6.58).

Install the Gloves

The gloves that we use on our large isolators (butyl rubber gloves; Figure 6.59) are different from the ones we use for the small isolator (butyl, dry box style, number 9). The butyl rubber gloves are sturdier than latex gloves and are preferred for the large isolators because they last longer and are less liable to damage. All of the components (three O-rings, cuff, glove) must soak in sterilant for at least

Figure 6.57 Cages and materials ready for autoclaving to be entered into a new breeding isolator.

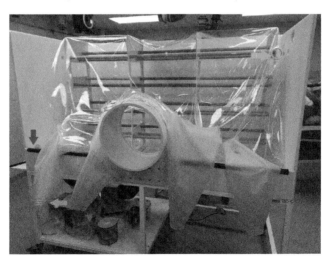

Figure 6.58 Large isolator ready for assembly. At this point, all isolator openings, including the port, the glove openings (orange arrow), the intake filter opening (blue arrow), and the exhaust filter opening (green arrow) are still open.

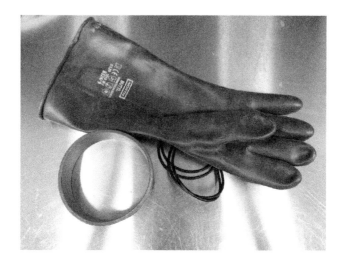

Figure 6.59 Components for installing gloves: gloves, three O-rings, and cuff.

45 minutes prior to assembly. For clarity, however, the illustrations discussed next show a dry procedure without the use of sterilant:

Procedure

1. Begin by inserting the cuff. The cuff is a gray, grooved plastic glove support (Figure 6.60) that should be inserted into the glove for approximately twice its width and fitted into the base of the glove (Figure 6.61).

Figure 6.60 Glove cuff.

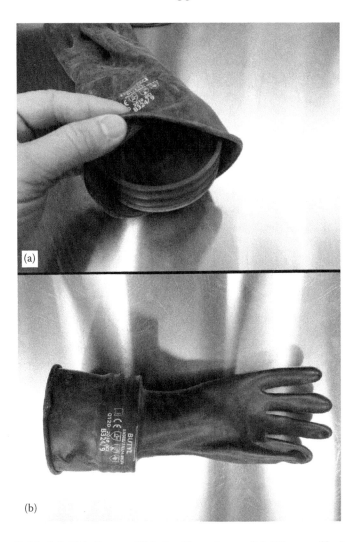

Figure 6.61 (a) Fit the cuff into the glove. (b) The cuff should be inserted into the glove for approximately twice its width.

2. The glove and cuff are inserted into the isolator glove port, with the hand of the glove directed toward the isolator and the thumb up (Figure 6.62). The insertion of the glove and cuff is a tight fit. Carefully stretch the polypropylene and make sure the glove port extends 2 cm past the last groove closest to the open end of the glove.

3. Place an O-ring into the last groove (closest to the opening of the glove), securing the end of the glove port to the glove and cuff (Figure 6.63).

Figure 6.62 The glove and cuff are inserted into the isolator glove port, with the hand of the glove directed toward the isolator and the thumb up.

Figure 6.63 Place an O-ring into the last groove (closest to the opening of the glove), securing the end of the glove port to the glove and cuff.

4. Fold the extra glove material over the end of the glove port and the cuff (Figure 6.64).

5. Place another (second) O-ring on the furthest groove of the cuff from the glove opening (Figure 6.65).

6. Fold the remaining glove material over the second O-ring. Place the third O-ring in the last groove in the middle (Figure 6.66). This O-ring can be placed next to the end of the glove material or under it.

7. During the steps described previously, the glove and port will be constantly wet from sterilant. Use isopropyl alcohol to dry the area before applying tape.

8. Use nylon tape to cover the glove material and edges of the cuff (Figure 6.67).

Figure 6.64 Fold the extra glove material over the end of the glove port and the cuff.

Figure 6.65 Place another (second) O-ring on the furthest groove of the cuff from the glove opening.

Figure 6.66 Fold the remaining glove material over the second O-ring. Place the third O-ring in the last groove in the middle.

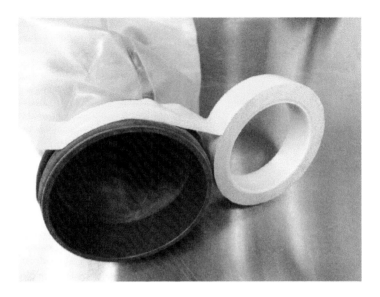

Figure 6.67 Use nylon tape to cover the glove material and edges of the cuff.

Assemble the Airflow Piping

1. After autoclaving (see previous discussion on preparing the fittings), place all the fittings in a sterilant bath for 20 minutes (Figure 6.68).

Figure 6.68 Soak the autoclaved pipe fittings in sterilant.

(a)

(b)

Figure 6.69 Two views of the bulkhead fitting: (a) taped and ready to install; (b) installed in the isolator housing (outside view).

2. Wipe the filter port openings and the surrounding area with sterilant. Install both bulkhead tank fittings (Figure 6.69) into each filter port opening. This step requires a tight seal. Use gaskets on each side of the fittings. Hand tighten the bulkhead fittings.

3. Hand screw in plugs inside the isolator (Figure 6.70).

4. Install the intake 90° assembly into each bulkhead fitting (Figures 6.71 and 6.72).

5. Fill the piping with sterilant (Figure 6.73). This will provide both a chemical barrier and a leak indicator later in the assembly process (Figure 6.74).

Note: This "double-barrier technique" is used in both filter changes (large isolators) and glove changes (see Chapter 10).

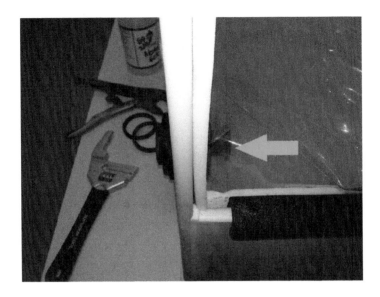

Figure 6.70 Interior, threaded plug (green arrow).

Figure 6.71 The 90° fitting assembly, sterilized and ready to be installed.

Figure 6.72 The 90° fitting assembly installed in the isolator housing.

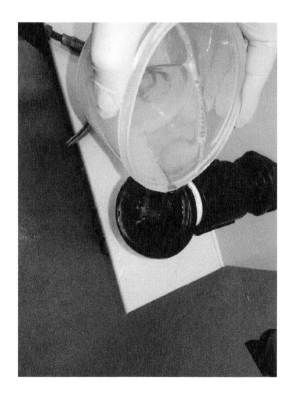

Figure 6.73 Fill fitting assembly with sterilant.

Figure 6.74 The 90° fitting assembly installed and filled with sterilant.

At this point, you should have the gloves installed, both bulkhead fittings installed in the filter ports, and the 90° assemblies installed in both filter ports (Figure 6.75).

Install the Intake and Exhaust Filters

1. It does not matter which filter is installed first.

   a. Installation of the exhaust filter is illustrated in Figure 6.76. The filter adapter is at the end of the previously installed filter assembly. The 90° elbow (yellow arrow) has an internal gasket between the female filter adapter (blue arrow) and the filter.

   b. Attach the steel clamp, which secures the two pieces without threads (purple arrow in Figure 6.76) to the end of the filter assembly,

   c. Fill the pipe assembly with sterilant again (Figure 6.77).

   d. Aseptically unwrap and screw in the exhaust filter (orange arrow in Figure 6.76). The female filter adapter should be filled to the brim with sterilant.

2. Install the intake filter: The same principles are repeated to install the intake filter. There are two differences between the intake and exhaust filters. The first is the intake filter housing, and the second is the final horizontal alignment of the intake filter. The housing forces airflow from the ventilation

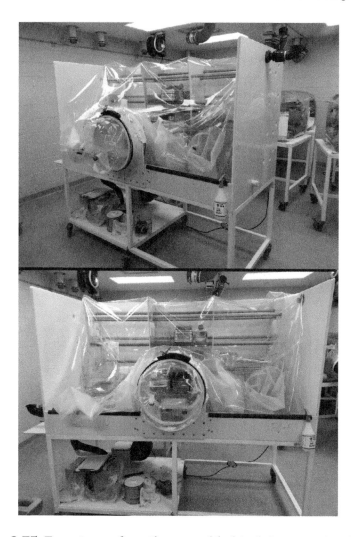

Figure 6.75 Two views of partly assembled isolator prior to stocking supplies.

motor through the intake filter. The intake filter will be vertical during installation (because of the chemical barrier) then placed horizontal after the sterilant is drained and the filter is aseptically installed.

a. Soak the bottom component of the intake filter housing for 20 minutes (Figure 6.78).

b. Screw the intake filter into the bottom portion of the intake filter housing (Figure 6.79).

Figure 6.76 Appearance of the filter assembly with filter installed: The filter (orange arrow) is installed onto the filter adapter (blue arrow) and secured with a clamp (purple arrow). The yellow arrow indicates the previously installed fitting assembly.

Figure 6.77 Fill the 90° fitting assembly with sterilant.

Figure 6.78 Bottom component of intake filter housing.

Figure 6.79 Intake filter housing screwed into the intake filter.

Figure 6.80 Intake filter and bottom component of filter housing installed in the 90° fitting assembly.

    c. Screw the filter/filter housing bottom (red arrow, Figure 6.80) into the polypropylene (black) fittings.

    d. Install the clear housing (Figure 6.81) and ensure that it is properly aligned.

    e. There are three screws that hold the top and bottom of the clear plexiglass. Hand tighten these screws once the housing is aligned (Figure 6.82). The intake filter is now installed (Figure 6.83).

    f. Both filters should be in place. All piping connecting the filters should still be filled with sterilant (to the level of the filters) and plugged.

Stock and Inflate the Isolator

This procedure is similar to the procedure for small isolators.

1. Place the autoclaved caging, sterilant, terrycloth towels, two inside caps, pens, rubber bands, protective mat, and water bottles inside the isolator.

2. Set the compressor to 110 psi. The higher pressure compared to the small isolator assembly saves time because it takes longer to pressurize the large isolators.

Figure 6.81 Place the filter housing.

Figure 6.82 Secure the filter housing with screws.

Figure 6.83 The filter housing is installed.

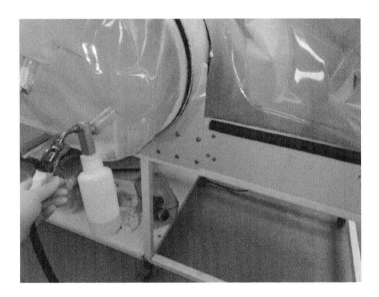

Figure 6.84 Inflate the isolator.

3. Remove one rubber stopper and inflate the isolator (Figure 6.84). As the isolator inflates, adjust the connection points (bulkheads, port, filter ports) if necessary. This will help correct the position of the canopy.

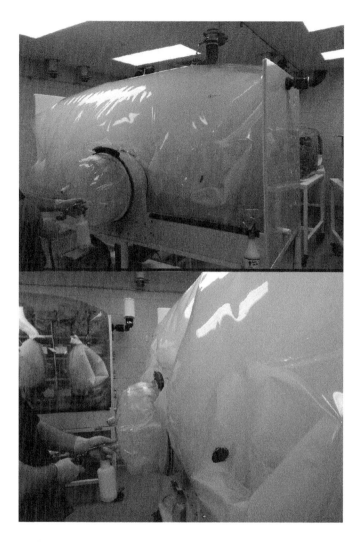

Figure 6.85 Adjust the canopy as the isolator inflates.

4. Inflation may take 20 minutes. During this time, sterilant vapor will slowly fill the interior of the isolator. Adjust the canopy manually as the isolator inflates (Figure 6.85).

5. After the isolator is fully inflated (Figure 6.86), plug the nipple on the cap and wait at least an hour. If the isolator deflates in that time, you will need to check the seal around the bulkheads, gloves, and port/canopy connections.

6. After an hour, the isolator should look like Figure 6.87.

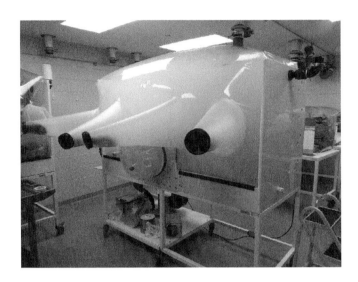

Figure 6.86 Fully inflated isolator.

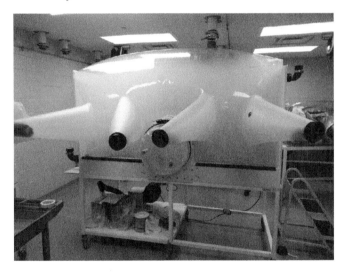

Figure 6.87 Fully inflated isolator after leak test.

Clean the Inside of the Isolator

This step is similar to the procedure for the small isolator assembly.

1. As described previously, partially deflate the isolator by aseptically removing one rubber stopper from the outer cap (Figure 6.88). Once the isolator gloves point directly to the ground (Figure 6.89), reinsert the rubber stopper.

Figure 6.88 Remove one rubber stopper.

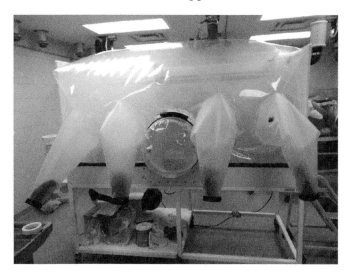

Figure 6.89 Partially inflated isolator.

Chlorine gas will exit the isolator at this time. Be cautious and make sure there is proper ventilation in the room.

2. Clean the exterior of the isolator with isopropyl alcohol. Carefully place all caging to one side of the isolator. Begin washing the interior of the empty side of the isolator with sterilant.

Every square millimeter of the interior must be physically scrubbed with sterilant and a terrycloth towel. This includes the mat, corners, and joints of the cage rack. Pay special attention to the cuffs of the gloves, filter ports, and port. Be aggressive but careful while washing. There are many sharp metal edges on the equipment. This step may take several hours, but it is crucial to be diligent.

3. After washing, wait 3 hours. Remove the filter port plugs from inside the isolator. Sterilant should pour out of the air supply fittings and into the isolator.
4. Tighten the hardware on the cage rack.
5. Wash the area of the filter ports thoroughly with a terrycloth towel.
6. Lay the intake filter housing horizontally (Figure 6.90) and connect it to the air hose, which is connected to the motor.
7. Connect the ventilation supply valve and hose with the ventilation motor and plug in the motor.

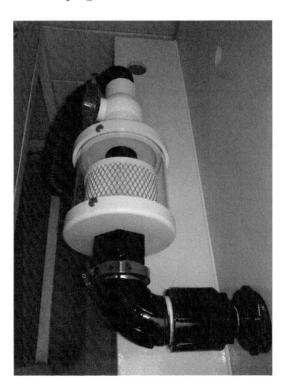

Figure 6.90 The final horizonal position of the intake filter connected to the air supply.

8. Make sure that the valve (if applicable) is open (i.e., parallel to the pipe).

Important note: This step **must** be done *after* removal of the filter port plugs and drainage of sterilant into the isolator. If it is done before you remove the plugs, sterilant will leak through the filter and dry there, potentially clogging it.

Start the Isolator

1. Make sure the motor is functioning properly. Confirm the positive pressure of the isolator. As the pressure builds, air will vent passively through the output filter.

2. To avoid damage to equipment, rapid drying is important. Chlorine dioxide is an oxidizer and will cause extreme rusting on metal (even stainless steel). To avoid this, position the caging equipment inside the isolator for optimal drying as follows:

   a. Place the cages upside down and use the wire on a cage as drying racks for smaller items like sipper tubes or cage cards.

   b. Hang the terrycloth towels.

   c. Use a small jack and slightly raise one side of the isolator. This will make excess sterilant collect at the other side of the isolator.

3. After 48 hours, soak the dry terrycloth towels in the pools of excess sterilant and wring them out into an empty cage. Keep repeating until the isolator is free of all excess sterilant.

4. Use sterile water to rinse the inside of the isolator (especially rubber bands). Use the empty water bottle to store the excess sterilant for later removal.

Prefilters

The use of a prefilter is strongly recommended. A prefilter is a thin piece of foam used to filter large dust particles (Figure 6.91). The prefilter is placed in the opening of the bulkhead fitting on the interior of the isolator. It is ETO sterilized and entered through the isolator port after the isolator is completely assembled. A prefilter will keep the exhaust HEPA filter clear of large debris. The use of a prefilter greatly extends the life of the HEPA filter.

Figure 6.91 Prefilter.

Port Entry and Exit

Introduction

All materials and animals must enter or exit the isolator through the port (Figure 7.1). The port is a double-barrier system consisting of two caps: the inside cap and the outside cap. The inside cap is solid (no nipples) and in our isolators is held onto the port by 18" × 1" rubber bands. The outside cap has two 1" nipples and is held on the port by a grooved gasket and an 18" clamp. We find the soft rubber band closure for the inside cap and the more sturdy clamp and gasket closure for the outside cap are the most convenient, but any secure method of closure that can be manipulated easily and not damage the equipment will work. The two nipples in the outside cap are for fogging and can also be used to introduce small objects into the isolator. They are sealed by 1" rubber plugs. To introduce or remove an animal or material into or out of the isolator, you must remove the outside cap, surface sterilize the interior of the port, introduce the materials, close the outside cap, fog the port, and then open the inside cap to enter the materials.

Supplies

- Respirator personal protective equipment (PPE)
- Sterilant
- Sterile terrycloth towels

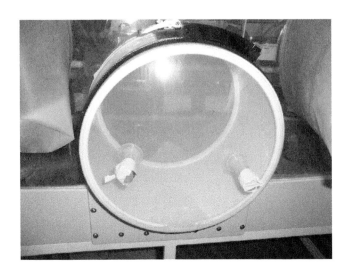

Figure 7.1 Isolator port.

- Atomizer
- Compressor (80 psi)
- Clean cart/table space
- Nylon tape
- Chemical respirator

Preparing, Opening, and Closing the Port

1. Spray isopropyl alcohol around the gasket and underneath the outside cap. This will make the next removal easier and result in less stress on the material.
2. Carefully remove the clamp and the grooved gasket from the outside cap.
3. Remove the rubber plugs from the outside cap and place them in sterilant to soak.
4. Remove the cap (Figure 7.2) and place it on a clean surface.
5. With a clean, presterilized terrycloth towel, wipe up any used sterilant remaining in the port from the previous entry (Figure 7.3). This will prevent any dilution of the newly applied chlorine dioxide. Discard the towel.
6. Soak a clean, presterilized towel in fresh chlorine dioxide. Starting in the center of the inside cap, gently scrub the area

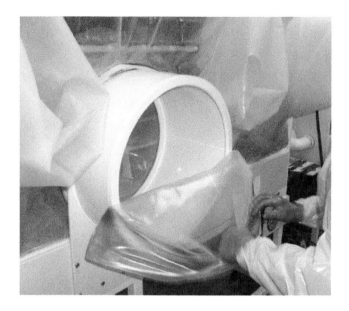

Figure 7.2 Remove the outside cap.

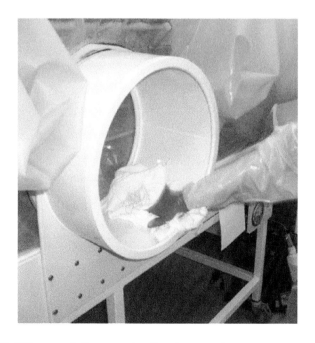

Figure 7.3 Wipe up left over sterilant in port.

Figure 7.4 Surface sterilize the port with sterilant, working from the center of the port to the outer rim.

Figure 7.5 Wash and inspect the outside cap.

in a circular pattern to the inside lip of the port. Use firm pressure and create a soapy lather. Continue to wipe down the inside circumference of the port. The last surface should be on the inner outside lip of the port and the external surface area (Figure 7.4).

7. Liberally wash all surfaces of the outside cap (Figure 7.5). Inspect the welds/seams of the plastic and carefully inspect the outside cap for damage.

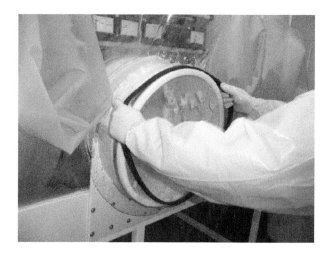

Figure 7.6 After sterile supplies have been placed into the sterilized port, replace the outer cap and gasket.

This is the point at which material or cages will be placed inside the port. Anything to be entered here must have been presterilized prior to entry (see Chapter 5 regarding sterilants and surface sterilization). It is then removed from its wrap or surface sterilized before placing it in the port and closing the outside cap (see the next section on proper placement of materials). *The outside cap must be replaced and the port fogged prior to opening the inside cap.*

8. Place the outside cap back on the port (Figure 7.6). Carefully place the grooved gasket around the port and ½" away from the outside lip.

9. Secure the 18" clamp in the grooved gasket and use the atomizer to fog the interior of the port (Figure 7.7). Note that both nipples are unplugged at this step.

10. Fogging is complete when the port is completely filled with chlorine dioxide vapor and "backflow" (escape of chlorine dioxide vapor) can be seen exiting the nipples.

11. Once the static air has been replaced with chlorine dioxide vapor, plug one nipple to pressurize the port.

12. Pressurize the port through the other nipple while allowing vapor to escape through the nipple for 30 seconds.

13. Quickly plug the second nipple with a 1" rubber stopper so that the port remains pressurized. The outside cap should be fully convex (Figure 7.8).

Figure 7.7 Fog the port.

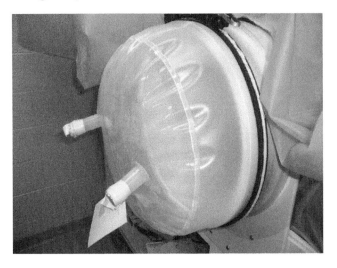

Figure 7.8 Following fogging, the outside cap should remain convex.

14. Secure the rubber stoppers with nylon tape (see Figure 7.8).

15. Note the amount of pressure within the port. After you allow a 60-minute contact time, check the pressure of the port and look at the outside cap to gauge whether pressure within the port has been maintained. If it has, the outside cap should be fully convex. Loss of the convex shape indicates loss of pressure and damage to either the inside cap or the outside cap.

If there is loss of pressure, remove the outside cap, and either

    a. Reinspect it or replace it and repeat port preparation *or*

    b. Replace the inside cap with the spare that is stored inside the isolator (see setup) and repeat port preparation.

(Remember not to open both the inside and the outside caps at the same time.)

16. If the port retains pressure (remains convex) after 60 minutes, it is safe to enter the isolator.

Proper Placement of Materials

Proper placement of sterilized items in a port is crucial for cold sterilization to work. As you place the items in the washed port, make sure the materials are not in direct contact with each other. Do not stack items against each other or stack boxes on each other, because direct contact will limit the chlorine dioxide exposure to the surface area of the materials. Figure 7.9 demonstrates the correct placement of the materials. Notice the open space. The water bottle is placed in the cage, and the lid is propped up on the side of the cage.

As you fog the interior of the port, notice the mist swirl in the crevices of the material. This will allow proper contact with the sterilant.

Figure 7.9 Proper placement of items in the port ensures optimal contact with vaporized sterilant.

Sterilizing Food and Supplies

Introduction

Materials may be entered into the isolators in several ways. Animals in cages, solitary items, and smaller items are entered directly through the ports (see Chapter 7). For entry of large items and supplies such as food and bedding, we use supply cylinders (Figure 8.1). These are filled and autoclaved to be aseptically placed into the isolator via the isolator port through a transfer sleeve. The contents of sterile cylinders may be deposited directly into breeding or experimental isolators. However, we prefer to place the contents first into supply isolators, which can then be monitored several times for sterility before moving the material into the breeding isolators. This practice greatly diminishes the chance of contaminating breeding isolators.

Sterility Indicators

Different types of indicators monitor different aspects of the sterilization process (see Figure 3.8). For example, autoclave tape, ethylene oxide (ETO) tape, and steam strips (see Chapter 3) indicate whether the appropriate conditions for sterilization have occurred (e.g., sufficient temperature or pressure). These are included in every autoclave and ETO load. They are helpful for determining if something has been autoclaved, but they do not tell us whether the object is actually

Figure 8.1 Supply cylinder assembled and ready to be filled for autoclaving.

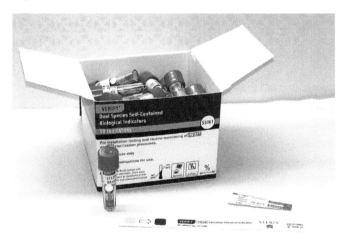

Figure 8.2 Steam strips and Verify® ampoule for inclusion in every supply cylinder.

sterile. Steraffirm® indicators (see Figure 3.8) are used to verify sterilization of water. Like autoclave tape, they change color (from red to green) when the correct temperature has been reached rather than indicating whether the load is sterile. Steraffirm® tubes measure two critical parameters: time and temperature (see "Water Sterilization" on page 122).

To determine if a load was indeed sterilized, we use self-contained biological indicators (SCBIs). The indicators that we use (Verify® ampoules; Figure 8.2) are included when autoclaving dry loads (such as the supply cylinders). These indicators should be placed in the densest part of the load prior to autoclaving. Unlike color-change indicators, Verify® ampoules indicate whether sterilization has been effective. They contain a disk inoculated with spores of two bacterial species. Within the plastic exterior portion of the tube, separated from the disk, there is a glass vial that contains soybean growth medium with a pH indicator. Prior to autoclaving, the color of the growth medium is blue. After exposure to heat, it will turn black. After autoclaving, the ampoule is retrieved, and the glass tube in the interior is cracked, exposing the inoculated disk to the growth medium. The ampoule is then incubated for 48 hours at 37°C. If the spores survived, the bacteria will grow, changing the pH of the medium. The medium will turn yellow, indicating failure of sterilization. The procedure for using the Verify indicator is as follows:

1. Retrieve the autoclaved ampoule from the sterilized load.
2. Crack the glass interior tube by applying pressure to the top of the ampoule (Figure 8.3).
3. Incubate the ampoule for 48 hours at 37°C (Figure 8.4).
4. If the ampoule turns yellow, discard the load.

Figure 8.3 Autoclaved ampoule ready to be cracked.

Figure 8.4 An incubation chamber designed for Verify ampoules.

Assembling Supply Cylinders: Installing HEPA Filtration Media

In our laboratory, the husbandry materials needed to maintain an axenic colony are sterilized in 18"-diameter slotted metal cylinders. These supply cylinders are hand wrapped with DW-4 HEPA filter material, similar to hand wrapping the isolator filters (see Chapter 6). This allows steam, heat, and pressure to penetrate the cylinder and its contents. There are several ways to pack the cylinder, but any method used must permit sufficient entry of heat, steam, and pressure to fully sterilize the contents. Therefore, packing correctly is vital. We have found the process that follows works well.

Supplies

- Regular personal protective equipment (PPE)
- Cylinder frame
- 24" × 128" DW-4 HEPA media
- T square and ruler
- Sharp razor blade
- Autoclave tape
- Mesh nylon laundry bag (large)

- Supplies for autoclaving (feed and bedding, paper bags, other supplies)
- Polyester film (Mylar®)
- Silicone gasket
- Rubber band
- High-temperature tape

Prepare the Cylinder

1. Start with a clean cylinder frame that has been washed and all adherent material removed. Use a clean and uncluttered workspace (Figure 8.5).
2. Measure the HEPA filter material. For our sterilization cylinders, we measure 24" × 128" DW-4 HEPA material.
3. Use a square to cut and a ruler to measure. Make sure the razor blade is new.
4. Straighten the HEPA material and place it into position. The filter material should overlap the edges of the filter ventilation slots by at least ½". Use a strip of autoclave tape so the material is tight against the filter (Figure 8.6).
5. Once the HEPA filter is in place, secure it using the 60" stainless steel clamp over the overlap area. Cover the clamp and the edge with nylon tape. Do the same for the other edge.

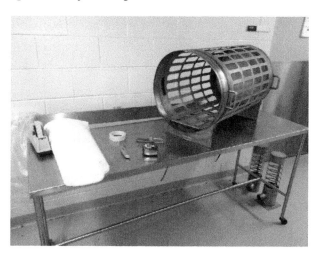

Figure 8.5 Cylinder frame, HEPA media, and materials for assembly.

Figure 8.6 Attach HEPA media to frame.

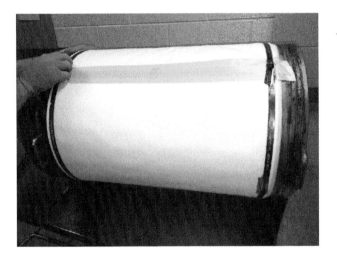

Figure 8.7 Cover frame and secure with nylon tape.

Use nylon tape to secure the edge of the HEPA material (Figure 8.7).

6. Completed assembly is shown in Figure 8.8.

Fill the Cylinder

1. First, place an empty nylon laundry bag within the cylinder (Figure 8.9). This will make emptying the cylinder easier.

Figure 8.8 Assembled cylinder.

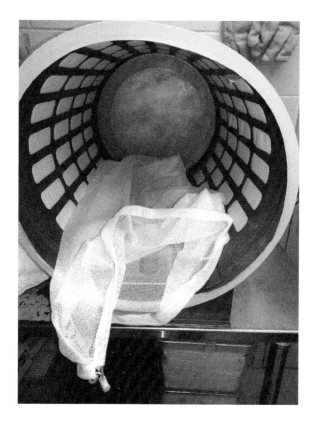

Figure 8.9 Place mesh bag in cylinder.

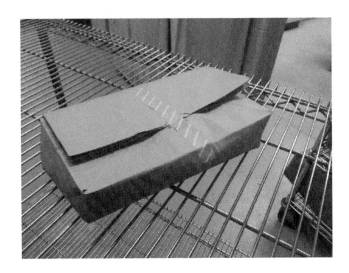

Figure 8.10 Food bags should be no more than 3" thick.

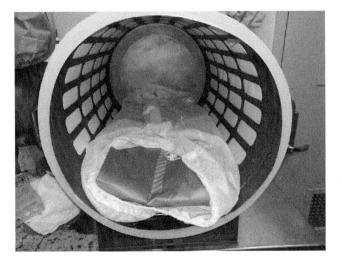

Figure 8.11 Place food bags flat in mesh bag.

2. We autoclave feed in a brown paper bag (12" × 18") filled to a thickness of no more than 3" (Figure 8.10). You can fit about four bags of food in one cylinder.

3. Place a biological indicator in the center of the middle bag.

4. Position the feed bags inside the laundry bag, flat on the grate (Figure 8.11).

5. We autoclave bedding in 12" × 18" brown paper bags. Bags with bedding should not exceed 6" in thickness (Figure 8.12).

Figure 8.12 Bedding bags should be no more than 6" thick.

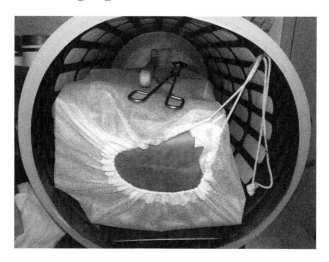

Figure 8.13 Place four bags of bedding on top of feed bags inside mesh bag.

6. Position four bags of bedding inside the laundry bag on top of the feed bags (Figure 8.13).

7. Secure the rope on the nylon laundry bag with a piece of autoclave tape.

8. A few small items can be placed on top of the supplies. If the cylinder is for initial isolator setup, items such as large crucible pliers, a long brush, ear tags, and the like can be added, but avoid overfilling the cylinder.

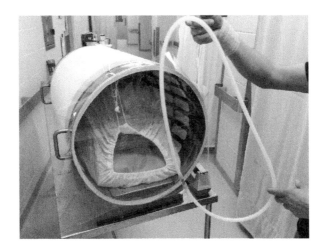

Figure 8.14 Polyester film sheet (Mylar) is placed over the cylinder opening, secured with tape, and held on with a silicone gasket.

Figure 8.15 A rubber band is placed over the gasket.

9. Place a Mylar sheet over the cylinder opening and secure it with small pieces of autoclave tape.

10. Place a loose gasket made of silicone tube over the square sheet (Figure 8.14).

11. Secure a tighter rubber band around the film (Figure 8.15).

12. The Mylar sheet must be centered and straight.

13. High temperature causes pinholes to form where the polyester film is overlapping itself (Figure 8.16). To prevent this, the edges of the film must be completely wrapped with tape.

Figure 8.16 The film overlaps itself around the edge of the cylinder. This is commonly where pinholes form.

14. Remove the silicone tube (Figure 8.17).

15. Starting at the edge, wrap high-temperature tape (see Chapter 3) once around the circumference of the cylinder (Figure 8.18).

16. Cut the tape with a sharp razor blade (Figure 8.19).

17. Pull the rubber band toward the back of the cylinder (Figure 8.20). The high-temperature tape should stay in place.

18. Cut off the free edge of the film (Figure 8.21). Starting at the edge, wrap the high-temperature tape around the circumference of the cylinder (Figure 8.22). Overlap the previous tape. Work your way to the back lip, then to the starting point of the edge.

19. Repeat, overlapping the existing tape (Figure 8.23).

20. Cut the tape at the top of the cylinder with a sharp razor blade.

21. Inspect the film seal for damage (Figure 8.24). Autoclave at 132°C for 25 minutes.

Figure 8.17 Remove the silicone band. Film is held on with the rubber band.

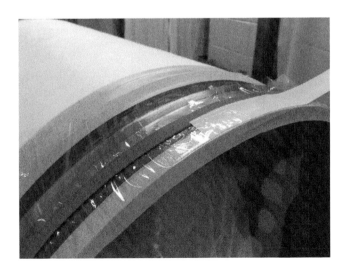

Figure 8.18 Secure the film with high-temperature tape. Start at the edge of the cylinder.

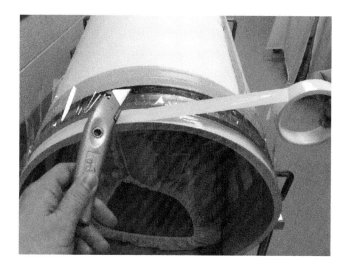

Figure 8.19 Cut the tape with a sharp blade.

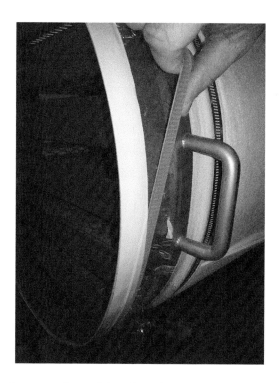

Figure 8.20 Remove the rubber band.

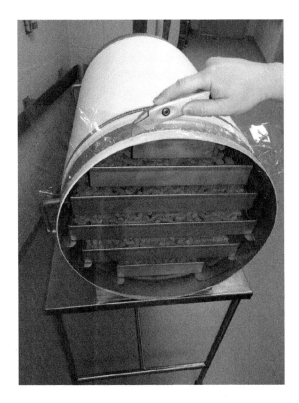

Figure 8.21 Cut off the excess film. *Note*: The cylinder in this photograph contains food trays rather than mixed supplies as described. Food trays can be used when food alone is to enter an isolator.

Water Sterilization

Water sterilization is a relatively simple procedure. We use 1- and 2-L water flasks (see sources, Appendix) to sterilize our water (Figure 8.25). Water that is treated either by reverse osmosis (RO) or by distillation is acceptable. Both processes result in water that is cleaner than tap water and contains limited amounts of minerals. The absence of minerals in the water prevents deposition of calcium residue with repeated use of the water bottle flasks.

We autoclave our water on a liquid cycle at 125°C for 60 minutes. Steam strips and SCBIs such as used in supply cylinders are not well suited for a liquid cycle. We use Steraffirm® Control Tube indicators (see Figure 8.25). These tubes detect two important factors: time and temperature. The tubes are placed inside a full water bottle. After an

Figure 8.22 Wrap the high-temperature tape around the circumference of the cylinder.

Figure 8.23 Overlap the existing tape.

Figure 8.24 Cylinder is sealed and ready to autoclave.

Figure 8.25 One and two-liter water bottles and Steraffirm indicators.

exposure of 125°C for 15 minutes, the tubes change color from red to green, indicating a successful cycle.

We use damaged 2-L water bottles as "testing bottles"; they are placed in the middle of a load, usually in the middle of the cart. A testing bottle is not used to supply water but is emptied after successful sterilization of the load.

Figure 8.26 Water bottles ready to fill.

Figure 8.27 Green caps soaking in warm water prior to use.

Supplies

- 1- and 2-L water bottles (Figure 8.26)
- Green rubber lids (Figure 8.27)
- Distilled or RO water
- Steraffirm indicators
- 5" × 8" container
- Dish soap

Figure 8.28 Fill 2-L water bottles only to 1800 mL.

Procedure

1. Wash the water bottle flasks with warm water and dish soap.

2. Soak the green caps in warm water for 20 minutes (see Figure 8.27). Hydrating the caps helps create a tight seal.

3. Fill each water flask to the 1800-mL mark (Figure 8.28). There is a risk of "boil over" if the water bottle is filled to or beyond the 2000-mL mark. Overfilling will cause a loss of about 400 mL of water after the cycle is completed.

4. Fit the hydrated green caps around the top lip of the water bottle. Be sure to place the caps evenly.

5. Place the filled water bottles on an autoclavable cart. Space them evenly and place the testing bottle in the center of the cart (Figure 8.29, green arrow).

6. Autoclave the cart on a liquid cycle at 125°C for 60 minutes.

7. Inspect the testing bottle(s). Be sure to verify the Steraffirm® indicator has turned green (Figure 8.30).

8. Let the water bottles cool overnight.

Figure 8.29 Placement of bottles on cart for autoclaving. Steraffirm®
indicator is in the center bottle (arrow).

Figure 8.30 Steraffirm® indicator should turn green after autoclaving
(left). Red (right) indicates failure of sterilization cycle.

9. Once they are cool, inspect every water bottle. There should
be an obvious dip in the cap (Figure 8.31). The dip indicates
that there is an effective seal. If the dip is too shallow, the cap
has dry rot, or there is noticeable debris, discard the bottle.

Wear gloves when handling the autoclaved bottles: This prevents
skin oils from contaminating the bottle surface and interfering with
surface sterilization later.

Figure 8.31 Sealing of sterilized bottles is indicated by a dip in the green cap (right). Absence of a dip (left) indicates seal failure.

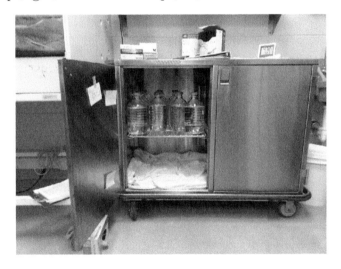

Figure 8.32 Bottles stored in a clean, protected location.

10. Store the sterile water bottles in a clean storage area (Figure 8.32). They will be surface sterilized before entry into isolators or hoods. Rotate the bottles and separate them by the date of sterilization.

Supply Cylinder Entry

Introduction

Autoclaved cylinders can be transferred directly into either large or small isolators or, as we prefer, into supply isolators. Supply isolators are used for holding the contents of two or three cylinders to verify the sterility of the contents prior to transfer into either experimental or breeding isolators (see Chapter 8). Supply isolators are useful in part because sterility indicators included in the cylinders are not always accurate and may not detect failure of steam to penetrate throughout all supplies. Also, they are not visible until after the cylinder is opened and the potentially contaminated material transferred into the isolator. For this reason, we test supply cylinders for bacterial contamination (see Chapter 15) three times at weekly intervals before transferring the contents.

Supplies are transferred from supply cylinders to isolators or between isolators using transfer sleeves (see Chapter 3). Transfer sleeves are polyurethane tubes of the same diameter as the isolator port. One end is connected to the port and the other to the cylinder or port of the supply isolator. We use cylinders of the same diameter as our isolator ports, allowing the transfer sleeve to be of constant diameter. However, step-down transfer sleeves can be made for transferring between ports of differing diameters. Transfer sleeves have nipples to allow inflation and fogging. Prior to transfer, it is essential that the junction between sleeve and port is airtight, and that the sleeve is completely inflated and without folds or wrinkles.

Supplies

- Respirator personal protective equipment (PPE)
- Sterilant
- Sterile terrycloth towels
- Atomizer
- Compressor (80 psi)
- Clean cart/table space
- Nylon tape
- Polypropylene connection sleeve (18")
- 2 rubber stoppers

Procedure

1. Make approximately 2000 mL of sterilant.
2. Wash the transfer sleeve and soak in the sterilant bath for 1 hour.
3. Remove the outside cap of the receiving isolator and prepare the port (Figure 9.1; see Chapter 7).
4. Carefully wash the cylinder seal and surrounding area (Figure 9.2). *Keep sterilant away from the HEPA filters.*

Figure 9.1 Remove the outside cap and prepare the port.

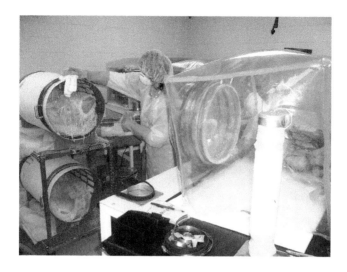

Figure 9.2 Wash the surface of the sterile supply cylinder.

Figure 9.3 Remove the transfer sleeve from the sterilant bath.

5. Examine the seal on the cylinder for leaks or tears.
6. Carefully remove the polypropylene transfer sleeve from the sterilant bath (Figure 9.3).
7. Place the gasket and gasket clamp on the isolator port prior to placing the transfer sleeve on the port.
8. Place the transfer sleeve on the port (Figure 9.4).
9. Clamp the gasket (Figure 9.5).

Figure 9.4 Place one end of the transfer sleeve over the isolator port.

10. Once the gasket and clamp are installed on the port, wash the seal again (Figure 9.6).

11. Place the cylinder gasket and clamp on the cylinder.

12. Carefully place the open end of the transfer sleeve on the cylinder (Figure 9.7).

13. Pull the sleeve back to the edge of the cylinder (cover the bare metal) (Figure 9.8).

14. Make sure the sleeve is even (Figure 9.9). Use the seam of the sleeve as a guide.

15. Fog the interior of the transfer sleeve (Figure 9.10).

Figure 9.5 Tighten the clamp.

Figure 9.6 Resterilize the cylinder seal.

16. Plug one of the nipples with a 1" rubber stopper.

17. Inflate the sleeve and pressurize until it feels firm (Figure 9.11).

18. If the sleeve is crooked, release the steel cable clamp, readjust one end of the sleeve, and reinflate. Alignment of the sleeve is important. A tight, straight sleeve makes it easier to transfer materials and to detect leaks.

19. Leave the pressurized transfer sleeve for 4 hours for the first transfer. For the second and third transfers, cylinder hookups require 1-hour contact times.

Figure 9.7 Place the other end of the transfer sleeve over the end of the cylinder.

20. While waiting, visually check the pressure of the sleeve. If there is a loss of pressure, inspect the seal of the cylinder for small holes or tears. Holes or tears will most likely be located around the edge of the Mylar® film.

21. After 4 hours of contact time, the cylinder is ready to be deposited into the isolator. This requires two people. One technician supports the cylinder (Figure 9.12), while another technician works inside the isolator.

22. Remove the inside cap and use large crucible pliers (3 feet) to break the cylinder seal. Poke a hole at 3 o'clock and move

Figure 9.8 Ensure that the sleeve reaches far enough over the cylinder and that it is straight.

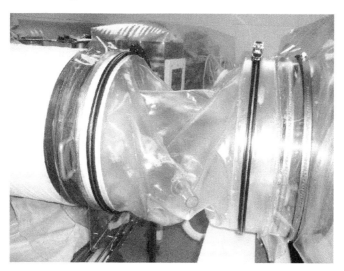

Figure 9.9 The transfer sleeve is clamped to the port on one end and the cylinder on the other end.

counterclockwise to 9 o'clock. This will leave a small amount of film at the bottom, preventing excess sterilant from dripping into the cylinder.

23. Pull the new supplies into the isolator (Figure 9.13).
24. Open all bags of food and place them in an empty container.

Figure 9.10 Inflate the sleeve.

Figure 9.11 The transfer sleeve is fully inflated and feels firm.

25. Create a "mold trap" with a variety of food pellets from each bag (Figure 9.14):

    a. Place the pellets in an empty container and add water.

    b. The mold trap must be wet at all times. This creates a favorable environment for mold or fungus, permitting early detection should contamination occur.

    c. This container of moist food can also be used for weanlings or sick mice if necessary.

Figure 9.12 One technician holds the cylinder outside the isolator while another technician working inside the isolator breaks the seal.

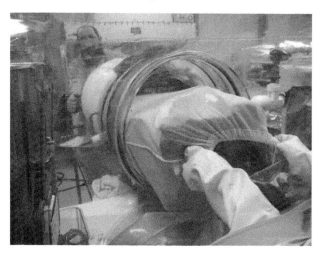

Figure 9.13 When the seal is broken, one technician pulls the mesh bag of supplies into the isolator, while another technician steadies the cylinder.

26. After the supplies are entered, place waste materials, empty water bottles, and anything else that needs to be removed from the isolator into the transfer sleeve.

27. Replace the inside cap.

28. Remove the transfer sleeve and the waste.

29. Close the port.

Figure 9.14 A mold trap consists of moistened chow. It is useful for early detection of fungal or bacterial contamination and can also be used to supply soft chow for weanlings, sick mice, or shipping isolators.

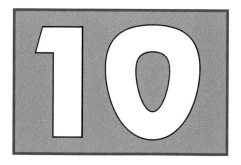

Isolator Maintenance

Introduction

We maintain individual isolators for varying periods of time. Small experimental isolators are kept running for 1–6 months or so, depending on the experiment. They are completely emptied and resterilized between uses. Breeding isolators can be kept going for 3 years or more. However, prolonged use requires consistent monitoring for damage and replacement of isolator components, such as filters and gloves. This chapter describes our practices for maintaining isolators for long periods. We also describe our methods for working in the isolators to maintain sterility.

Part 1: Cleaning and Maintenance

As mentioned, we use chlorine dioxide sterilant for surface sterilization. Chlorine dioxide has largely replaced peracetic acid for fogging isolators and cold-sterilizing equipment. It is relatively nontoxic, effective, and easy to use. However, it does have some disadvantages.

When chlorine dioxide dries, it produces a sticky residue. This residue can be an advantage when connecting filters to a canopy or any other connection. The sticky residue will help seal the connection by acting as a light adhesive. The disadvantage to this chemical is that it is corrosive and must be constantly removed from surfaces to prevent rusting and damage to equipment. If not removed, chlorine dioxide residue will eventually degrade metal parts, such as

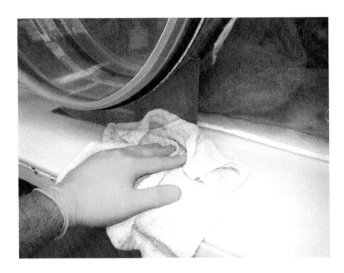

Figure 10.1 Wipe up excess chlorine dioxide around the port.

biosafety cabinets and metal tools. Therefore, chlorine dioxide should be removed as soon as possible after use.

For removal of chlorine dioxide, we use 60% isopropyl alcohol. For example, when we open an isolator port, we remove excess chlorine dioxide that remains from the previous entry (Figure 10.1) and then use alcohol to wipe down the area at the base of the port. In addition, after each port entry or exit, we wipe up the excess chlorine dioxide and clean the area by spraying the clamp, gaskets, port, table, and outside cap with 60% isopropyl alcohol (Figure 10.2). Keeping the area around the port clean is extremely important and will prevent future isolator contaminations.

In addition to cleaning and removing chlorine dioxide residue, there are many other useful applications for 60% isopropyl alcohol. Alcohol is a weak antiseptic, but it does remove and kill some microorganisms. Because it penetrates skin oils, it is a useful decontaminant for hands and contact surfaces. We use isopropanol to disinfect areas on the canopy that have frequent skin contact, such as areas of the canopy that contact the face and chest of workers (Figure 10.3). Isopropanol is also useful during manipulation within isolators. After donning regular personal protective equipment (PPE), we always liberally spray our gloved hands and arms before placing them in isolator gloves (Figure 10.4). This has two effects: It helps lubricate the gloves and sleeves, making it easier to access the isolator, and it decontaminates the interior of the isolator glove to some

Figure 10.2 Spray isopropyl alcohol to remove chlorine dioxide residue.

Figure 10.3 Use isopropyl alcohol to clean the outer surface of the isolator.

extent. If there is a small pinhole in the glove, the alcohol spray may to some extent prevent entry of bacteria from clothing or hands into the isolator.

Finally, alcohol is helpful for cleaning and maintaining the atomizer. The atomizer is designed to produce a fine mist, which is essential for proper contact of sterilant with surfaces. However, atomizers contain

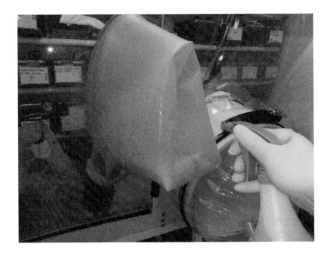

Figure 10.4 Use isopropyl alcohol to clean the inside of the glove sleeve.

seals and gaskets that are susceptible to degradation, especially with constant contact with chlorine dioxide. For this reason, at the end of the day's activities, we flush the atomizer with 60% isopropyl alcohol. Flushing the atomizer at the end of every day greatly extends the life of the tool.

Part 2: Replacing Gloves on an Active Isolator

Isolators can be constructed with gloves attached in various ways. In our large breeding isolators, butyl rubber gloves are attached at the ends of glove sleeves that are connected to the isolator canopy. These gloves can be replaced if necessary without compromising the isolator. In our small isolators, the glove and sleeve are a single unit that is attached directly to the canopy. These isolators must be resterilized to replace gloves.

Gloves should be inspected at least once a month. You can inspect a glove by putting a hand in the glove and then removing it while pulling the sleeve and glove inside out. Once the glove is inside/out, examine it visually for pinholes or other damage. It can be difficult to tell if there is indeed a small tear or hole. If you suspect a small hole but cannot see one, you can confirm it by pouring sterilant into the glove, applying a small amount of pressure, and waiting to see if sterilant leaks into the interior of the isolator. If there is damage, replace the glove.

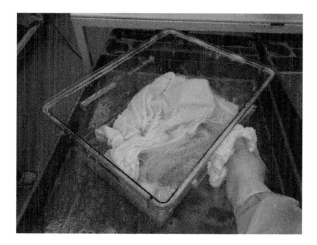

Figure 10.5 Sterilant for changing a damaged glove.

Procedure

1. Start by making 3 L of sterilant (Figure 10.5). Soak the replacement butyl rubber glove in the sterilant bath.

2. When replacing gloves or filters on an active isolator, you will need to provide two barriers: one physical and the other chemical. In the case of gloves, the physical barrier is a clamp that prevents contact between the interior of the glove and the interior of the isolator (Figure 10.6).

3. Place the clamp on the sleeve. Position the clamp as straight as possible. Make sure the ends are hand tightened as much as possible. If there is a visible hole in the finger, use nylon tape to seal the damaged area during manipulations.

4. The chemical barrier is sterilant that is introduced into the isolator. To enter sterilant into the isolator, first empty a sterilized bottle of water outside the isolator. Place the empty bottle and the green top in the sterilant bath with the replacement glove. Wash the bottle and top with a terrycloth towel.

5. Fill the bottle with approximately 1500 mL of sterilant and seal it with the green lid.

6. Prepare the isolator port (see Chapter 7) and scrub the water bottle with sterilant. Place the bottle in the port. Allow a 60-minute contact time.

7. Open the inside cap and the water bottle containing the 1500 mL of sterilant into the isolator.

8. Remove the lid and pour the sterilant in the sleeve (Figure 10.7).

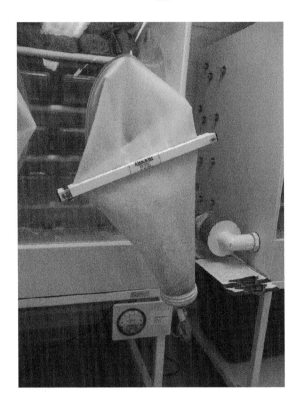

Figure 10.6 The clamp has been placed on the sleeve and the glove taped off with nylon tape.

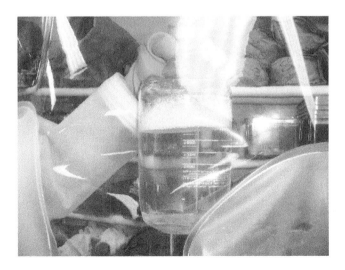

Figure 10.7 From inside the isolator, pour the sterilant into the sleeve.

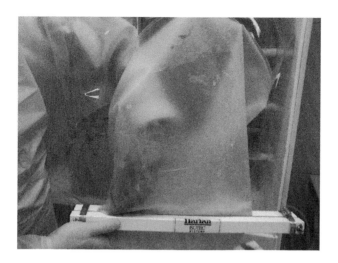

Figure 10.8 Ensure that the sterilant reaches all surfaces.

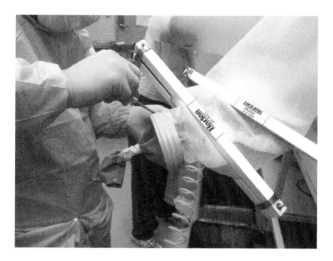

Figure 10.9 Two clamps may be necessary to prevent leakage.

9. Cover the entire area with sterilant and make sure that it reaches every crevice around the clamp (Figure 10.8).

10. Sterilant will leak past the clamps. Shake the sleeve so that the sterilant lathers and covers all the interior surface area. Use two clamps if necessary to contain the sterilant within the sleeve portion of the canopy (Figure 10.9).

11. Remove the glove assembly (cuff and O-rings; Figure 10.10). Once the glove is removed, wash and scrub the exterior of the glove port (Figure 10.11).

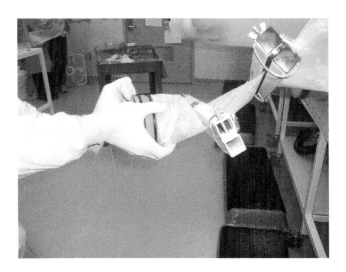

Figure 10.10 Remove the old glove.

Figure 10.11 Appearance of the clamped sleeve with glove removed. Surface sterilize the surrounding area prior to replacing the glove.

12. Replace the glove with the new, soaked glove from step 1 (see "Glove Assembly" in Chapter 6, part 2).

13. Remove the clamps and allow the excess sterilant to sit in the interior of the new glove and sleeve for 120 minutes. Drain the excess sterilant back into the empty water bottle inside the isolator.

14. Remove the water bottle in the next port entry.

Part 3: Repairing Holes in Isolators

Holes in the isolator canopy, while sometimes disastrous, do not always result in contamination of the isolator. In fact, it is largely the positive air pressure that protects the isolator from entry of microbes from many sources, and this positive pressure can protect the isolator in the event of small holes in the canopy or even seam failures. Thus, holes can sometimes be patched and the isolator saved. That said, contamination can occur, and the hole must be discovered quickly and patched before that happens. Time is critical, and action must be taken immediately. Two people are needed to patch holes.

If you find a hole, seam failure, or tear in the isolator canopy, proceed as follows:

1. Make 2 L of sterilant.
2. Soak an autoclaved terrycloth towel in sterilant and hold the soaked towel against the seam failure/tear. Keep holding the towel so it covers the entire area. Have one arm inside the isolator pressing against the other hand holding the towel on the outside. It is okay to have sterilant leak into the isolator.
3. Outside the isolator, empty a bottle of sterilized water.
4. Soak the empty water bottle and lid in sterilant.
5. Fill the bottle with sterilant, place the green lid on the water bottle, and wash the exterior with sterilant.
6. Introduce the bottle of sterilant and a sterilant-soaked terrycloth towel into the port. Follow appropriate port entry procedures (see Chapter 7). Because time is of the essence, decrease the contact time to 20 minutes.
7. Open the inside cap and retrieve the towel and bottle of sterilant into the isolator.
8. Take the soaked terrycloth towel and begin washing the entire interior area near the hole.
9. Place the soaked towel over the hole on the inside. You can now remove the sterilant-soaked towel on the exterior.
10. On the outside, spray 60% isopropyl alcohol over the area of the hole to dry it.
11. Take a 3.5-g packet of two-part urethane (Figure 10.12) and mix it using a Popsicle stick.

Figure 10.12 Urethane patching material.

Figure 10.13 Completed patch.

12. Cut a small piece of extra polyurethane (canopy material) to patch the hole. Make the patch at least an inch larger than the diameter of the hole.

13. Remove the soaked towel on the interior.

14. Make sure the area around the seam failure/hole is dry.

15. Spread the mixed urethane on the clean polyurethane patch.

16. Place the patch over the seam failure/hole.

17. Spread the remaining urethane over the edge of the patch (Figure 10.13).

18. Wait 24 hours.

19. Spread more urethane over the edge of the patch.

Part 4: Working in the Isolator

In this section we offer a few ideas for accomplishing common experimental tasks that can be difficult when working in an isolator:

1. Labeling samples
2. Weighing mice

Sample and Tube Labeling Inside the Isolator

Samples collected inside the isolator must be placed in sealed tubes and labeled aseptically in such a way that it prevents removal of the label during passage through sterilizing solutions when the sample is removed from the isolators. Sterilized markers are available (see the Appendix for sources) and work well for writing on paper and temporarily marking mice, but the ink washes off in sterilant, and permanent markers that do not wash off are not sterile. For this reason, labeling samples for removal from the isolator can be challenging.

Whenever possible, we use sterile paper strips to identify samples. Sterile paper can be entered into the isolator in cylinders or at setup. Alternatively, sterile cage cards can be used to label samples. Using a sterile pen, write down the sample ID on a cage card, cut the written sample ID off the cage card into a small strip, and place the strip into the tube with the sample (Figure 10.14). Seal the tube with the paper label. The paper is sterile and remains protected in the sealed tube during the process of removal from the isolator and transport to the laboratory. This process can only be used with dry samples (e.g., bedding, mouse tail samples, etc.).

Weighing Mice

Options for Weighing Mice

Experimental protocols often involve periodic weighing of individual mice. This requires a scale that can be sterilized, is accurate to at least 0.5 g, and can withstand manipulations inside the isolator. We have used digital scales, balance beams, and spring scales. A discussion follows of the strengths and weaknesses of the different types of scales.

Figure 10.14 This vial inside the isolator contains both the sample and a small strip of paper identifying the sample. Anything written on the outside of the vial will wash off in sterilant, but the paper remains dry and legible.

Digital scales can be sterilized by ethylene oxide (ETO) and entered into isolators. They are expensive, however, difficult to clean between uses, and easily damaged. We have not found them to be sufficiently sturdy to withstand isolator use. Triple-beam scales can be ETO sterilized or autoclaved and entered into isolators, and they can be surface sterilized and used in biosafety cabinets as well. They are more durable than digital scales but are subject to corrosion from contact with sterilant, and they can be difficult to calibrate and inconvenient to use. Nevertheless, they are practical for many uses and can be accurate if used with care.

A simple way to weigh mice is the use of a chemistry stand, disposable paper funnel, and a spring-loaded Pesola scale (Figure 10.15) (see the Appendix for sources). Pesola scales are relatively inexpensive, can be autoclaved (121°C), are small, and are easy to calibrate and use. They can be aseptically introduced into either isolators or class II biosafety cabinets (see Chapter 12). They are accurate only to 0.2 to 0.5 g, depending on the scale, and thus are less sensitive than either digital or triple-beam scales, but for most uses, their convenience outweighs any disadvantages.

Use of the Pesola Scale for Weighing Mice Aseptically

We use an autoclaved chemistry stand to hold the scale and autoclaved paper cones to hold the mouse for weighing. Wrap the

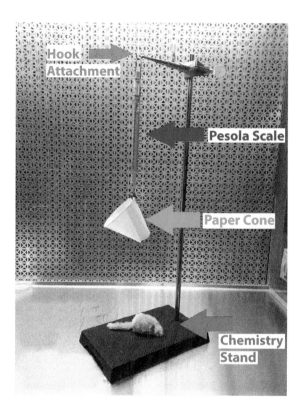

Figure 10.15 Pesola scale assembly with the parts labeled. The part of the mouse is played by a stuffed toy.

chemistry stand in a blue wrap and autoclave at a minimum of 121°C for 25 minutes. Place the paper cones and Pesola scales in peel pouches and autoclave at 121°C for 25 minutes (Figure 10.16). To weigh the mice:

1. Aseptically introduce the items in a sterile biosafety cabinet or isolator (Figure 10.17).
2. Use a sterilized hook clamp to hang the Pesola scale from the chemistry stand.
3. Open the peel pouch containing the paper cones and clip a cone to the Pesola scale (Figure 10.18). Tare the Pesola scale by turning the dial at the top of the scale to 0.
4. Gently place the mouse in the cone (Figure 10.19). The mouse will naturally want to "burrow" into the cone. This will keep the mouse busy as you weigh it. If you are patient and calm, the mouse will be calm as well.

Figure 10.16 Sterile chemistry stand, cones, and scale for weighing mice.

Figure 10.17 Pesola scale on stand in a biosafety cabinet ready to weigh mice.

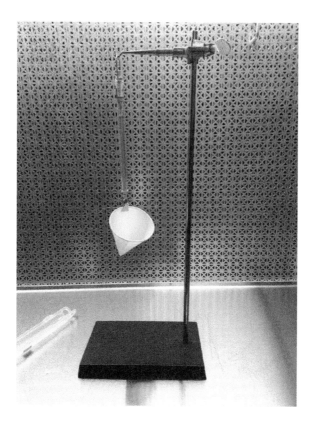

Figure 10.18 The weighing cone is clipped to the scale.

Figure 10.19 The mouse is placed in the cone and allowed to burrow into the tip. Cutting a small hole in the tip may encourage burrowing.

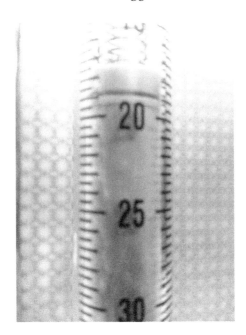

Figure 10.20 The scale is clearly marked in 0.5-g increments.

5. Once the mouse is calm, note the weight on the scale (Figure 10.20).

6. The scale can be reused within the same experimental group. If there are multiple groups within the biosafety cabinet, use a new, sterile scale for each group.

Aseptic Mouse Transfer

Introduction

There are two ways of transferring axenic mice from within an isolator to another location. External transfer means the mice will leave the confines of the germ-free barrier completely. The mice will be transported while inside a loosely sealed cage with a solid top and placed in either a class II biological safety cabinet or an individual Isocage or transferred to the laboratory for necropsy.

Internal transfer means the mice remain germ-free but are transferred to a different isolator or shipping container. This transfer uses the same principles as the cylinder deposition (Chapter 9).

Part 1: Internal Transfer between isolators

Introduction

An internal transfer refers to transferring mice between sterile isolators. The procedure is similar to those discussed in the chapters on supply cylinder deposition and shipping mice. Internal transfer is the safest way to transfer axenic mice. The most common reason to do an internal transfer is to expand an established breeding colony into a new, verified isolator.

An internal transfer of mice between two isolators requires two people: one handling each isolator. The process involves providing a passageway by attaching a transfer sleeve to each isolator port, sterilizing the connections, transferring the mice, and resterilizing and closing the ports.

Supplies

- Polycarbonate solid top
- Polycarbonate cage
- Respirator personal protective equipment (PPE)
- Sterilant
- Sterile terrycloth towels
- Atomizer
- Compressor (80 psi)
- Clean cart/table space
- Nylon tape
- Pesticide sprayer

Procedure

1. Make approximately 2 L of sterilant and soak the polypropylene transfer sleeve for 60 minutes. We use 18"-to-18" transfer sleeves because all of our cylinders and isolator ports have 18" diameters. Transfer between isolators with different diameters will require a step-down sleeve.

2. Wash the sleeve and inspect the material for any damage. Prepare the port (see Chapter 7) on the donor isolator and attach the sleeve to the port (see Chapter 9). Repeat the same steps on the receiving isolator.

3. Set the compressor to 80 psi. Inflate the sleeve. Using the seam as a guide, make sure the sleeve is straight. Be careful not to overpressurize the sleeve. If you do, the inside cap may blow off.

4. Allow 4 hours of contact time. Remove the inside cap of the donor isolator. Place the designated cage of mice inside the port. For internal transfer, no solid top is needed because the mice remain within the sterile isolator system. Close the inside cap of the donor isolator before the other technician removes the inside cap of the receiving isolator.

5. Closing the inside cap of the donor isolator first will protect the donor isolator colony in the established isolator from any possible contaminants in the newly established receiving isolator. Even though the new isolator has been microbiologically tested three times prior to the transfer, for maximum protection against contamination it is best to err on the side of caution.

6. The technician in the receiving isolator can now open the inside cap, retrieve the mice, and close the inside cap.

7. Once the inside caps are secured, you may detach the transfer sleeve, resterilize each port (see Chapter 7), and replace the outside caps.

Part 2: External Transfer of Mice

You can transfer mice externally for a number of reasons: to move from isolator to experimental hood, isolator to Isocage, or isolator to laboratory. Once mice leave their colony (isolator of origin), they never go back. The only exception is embryo transplantation surgery (rederivation), for which recipient females are removed for surgery and then replaced into a quarantine isolator (see Chapter 14).

Procedure

1. Sterilize the class II biosafety cabinet (Figure 11.1):

   a. This requires respiratory PPE (see Chapter 4).

   b. Clean the cabinet surfaces with a sterile terrycloth towel and sterilant.

   c. Spray the inside of the hood with sterilant after you physically wipe the interior. For this, the atomizer is not necessary. You can use a hand sprayer or a pesticide sprayer.

2. Prepare the port as usual (see Chapter 7).

3. Assemble the appropriate number of presterilized empty cages and solid tops to be entered into the isolator (see Figure 3.5B). Solid tops are used to prevent contact between mice and sterilant during surface sterilization of the cage for transfer. Empty cages may be needed to replace the cages being removed.

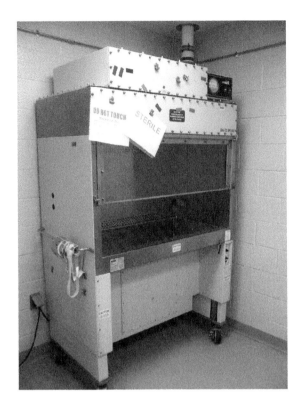

Figure 11.1 Class II biosafety cabinet, clean and ready to be surface sterilized for temporary housing of gnotobiotic mice (see Chapter 12).

4. Prepare the port for entry as described in Chapter 7.

5. Aseptically, place the cages and tops within the port, close the port, and fog.

6. Allow 1 hour of contact time. Use this time to prepare the mice and samples within the isolator.

7. Prepare the cages to be removed from the isolator.

8. Remove the wire, food, and water bottle from the cage inside the isolator.

   a. Take the identification card out of the cage card holder and place it inside the cage.

   b. Place the solid top on the cage and enter it in the port.

   c. Remove the cage from the port (Figure 11.2).

9. If the cage is to be transferred to a biosafety cabinet, it is placed in the cabinet and surface sterilized again before the solid top is replaced with a microisolator top (Figure 11.3).

Figure 11.2 Mice are removed from the isolator in cages with solid tops.

Figure 11.3 Solid-top cages may be placed in the sterilized biosafety cabinet and resterilized prior to replacing the solid top with a micro-isolator top.

Note: The solid top limits airflow from the outside. Mice can only be kept in solid-top cages for short periods, so work quickly.

10. Close and sterilize the port as described in Chapter 7.

11. If mice are to be transferred to a biosafety cabinet for short-term experiments, they must first be placed in new sterile cages in the appropriate experimental groups. Manipulation of mice in biosafety cabinets is discussed in Chapter 12.

12

Working with Germ-Free or Gnotobiotic Mice in the Class II Biosafety Cabinet

Introduction

As noted previously, we sometimes house gnotobiotic mice for short durations in class II biosafety cabinets. The class II biosafety cabinet performs two important functions: It protects the contents of the cabinet from outside contamination, while also protecting the user from infectious agents inside the cabinet. This housing is not germ-free, however, and keeping mice clean in this environment requires strict attention to detail. Strict aseptic technique is needed when handling sterile items, entering materials, or manipulating objects/mice in a class II biosafety cabinet. Also, because the only barrier between the inside of the cabinet and the outside is an airflow barrier, it is essential to take care not to interrupt this barrier any more than absolutely necessary. The airflow is disrupted every time an arm reaches into the cabinet or any time an object or body part obstructs the inflow or outflow zones. The airflow can be interrupted even when someone walks by the cabinet. For this reason, mice housed in biosafety cabinets are easily contaminated, particularly with bacterial spores or other airborne contaminants. That said, the advantage of the biosafety cabinets is that mice are easier to manipulate than in the isolators, and it is easier to prevent cross contamination between cages in the same biosafety cabinet than between cages in the same isolator.

To work in the biosafety cabinets you need sterile personal protective equipment (PPE) (see Chapter 4) and two people working together: one person (the primary operator) in sterile PPE working inside the cabinet and the other person (assistant) in standard PPE. Once the primary operator has donned sterile PPE, he or she can no longer come in physical contact with items outside the hood. The role of the assistant is to supply sterile materials and to remove used items.

Supplies

- Sterile PPE
- Class II biosafety cabinet
- Autoclaved, double-wrapped cage kits (Chapter 3)
- Other sterile items, depending on the experimental protocol

Procedure

Prior to transfer, the assistant should gather enough autoclaved, wrapped cage kits (see Chapter 3) to house the appropriate groups of mice, depending on the experimental design. The protocol for entering cage kits into the biosafety cabinet is as follows:

1. The assistant breaks the autoclave tape seal and removes the outer wrap (Figure 12.1).

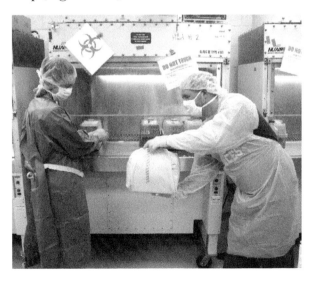

Figure 12.1 The nonsterile assistant removes the outer wrap.

Note: Do not touch the sterile inner blue wrap.

   a. Unfold the first corner to the left (Figure 12.2).

   b. Unfold the top and carefully pull the bottom corner toward you (Figure 12.3).

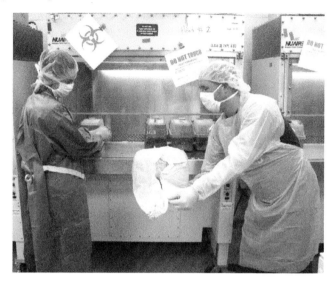

Figure 12.2 The nonsterile assistant unfolds the left corner of the wrap.

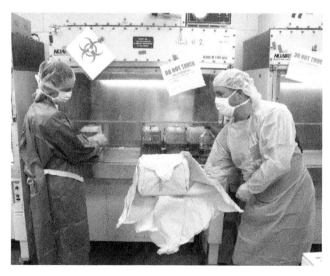

Figure 12.3 The nonsterile assistant carefully unfolds the wrap without touching the inner wrap.

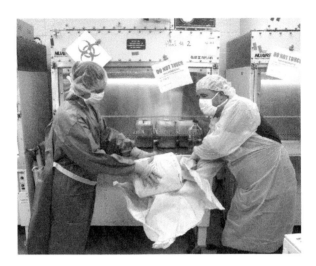

Figure 12.4 The sterile technician picks up the cage in its sterile inner wrap.

    c. Pull the last corner to the side. The exposed sterile inner layer is then accessible by the manipulator in sterile PPE. Remember, the outside blue wrap is not sterile and can only be contacted by the assistant.

  2. The manipulator can now pick up the sterile, wrapped cage (Figure 12.4), place it in the hood, and transfer the mice into clean, sterile cages (Figure 12.5).

After transfer, the cages are labeled using a sterile pen. Make sure that the mice have adequate water, food, and clean bedding (Figure 12.6). Double-check each cage for water, food, properly placed lid, damage to caging, proper cabinet flow/power, and placement within the biosafety cabinet.

Experimental Manipulations in the Biosafety Cabinet

Introduction

Preventing environmental contamination of mice in biosafety cabinets as well as preventing cross contamination between cages within the cabinet are challenging and require strict asepsis and attention to detail. Cross contamination can occur by contact of any object with

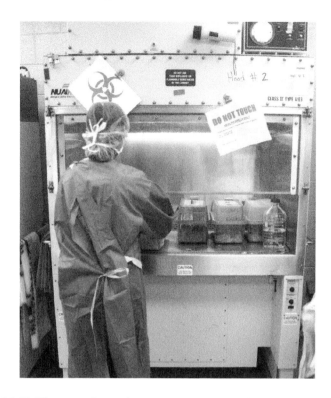

Figure 12.5 The sterile technician can now place the cage into the hood, remove the inner wrap, and transfer the mice to new cages.

Figure 12.6 Make sure that mice have adequate supplies, that cages are appropriately assembled, and that cages are spaced out and not blocking the hood airflow patterns.

Figure 12.7 Example experiment with three groups. Groups 1 and 2 are monocolonized with one organism each. Group C is the control (germ-free). The parts of the mice are played by pink, blue, purple, and green understudies.

the inside of a cage; by feces or urine contamination during handling of mice; spillage of water, food, or bedding; and so on. The description that follows is an example of the procedures we use to minimize the chance of cross contamination among cages colonized with different bacteria. In this example, there are three experimental groups in cages that have already been placed inside the cabinet. Each group is in its own cage. Group C is the control. Groups 1 and 2 are colonized with different organisms (Figure 12.7).

Procedures

1. Surface sterilize the area surrounding the boxes inside the cabinet (Figure 12.8).
   a. Without reaching into the cabinet or touching anything, liberally spray the sterilant between boxes and around the lower half of the cabinet.
   b. Avoid spraying the microisolator filter.
   c. Wait 5 minutes.
2. Gown into sterile PPE (see Chapter 4).
3. Manipulate group C and group 1:

Figure 12.8 The assistant surface sterilizes the inside of the biosafety cabinet prior to manipulations.

Figure 12.9 Group C (the control group) is handled first.

a. Manipulations within the cabinet may include husbandry (cage, food, or water changes), experimental procedures, or sample collection.

b. *The control (germ-free) group is always manipulated first* (Figure 12.9). This minimizes the chance of inadvertent contamination of the controls. Once the primary operator is properly gowned in sterile PPE, he or she may only touch sterile objects on or within group C (including animals).

Figure 12.10 Place a sterilant-soaked towel inside the biosafety cabinet.

   c. After the required tasks are completed in the control group, the assistant should spray the primary operator's gloved hands with sterilant. The assistant should also introduce a presterilized towel soaked in sterilant into the biosafety cabinet (Figure 12.10). This can be used to clean up any spills (Figure 12.11). The primary operator can then carefully open the cage and manipulate objects and animals within group 1.

Note: Animals can be very unpredictable. Use careful observation when opening a cage to avoid escape of mice or spillage of materials. Bedding, food, or other caging material can be dropped outside the microisolator unit. Immediately clean up any dropped material with the disposable towel soaked with sterilant (Figure 12.11). Dispose of the towel outside the cabinet. The assistant should replace the towel with another sterilant-soaked towel immediately.

  4. Manipulate the next experimental group:

   a. Once all of the tasks are completed within group 1, the operator exits the cabinet and removes sterile PPE. A change of PPE between colonized groups helps minimize cross contamination. Dispose of the sterile PPE in the biohazard trash (Figure 12.12).

   b. Spray the surrounding area inside the cabinet with sterilant. Avoid spraying the filter tops of the cages.

Figure 12.11 Use the sterilant-soaked towel to wipe the surface of the cabinet and clean up any spills prior to starting group 1.

5. Gown up in sterile PPE a second time. The assistant should introduce a fresh disposable towel soaked in sterilant (Figure 12.13).

6. Perform all of the tasks required in group 2. Once all of the tasks in group 2 are completed, wipe the area down using the disposal towel soaked with sterilant.

7. Exit the biosafety cabinet and properly dispose of your PPE in the biohazard trash.

8. Repeat steps 4–7 as necessary for any additional groups.

Figure 12.12 Discard sterile PPE and regown prior to working with the next colonized group.

Figure 12.13 Use a fresh sterilant-soaked towel and fresh sterile PPE for each group.

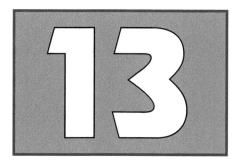

Shipping Mice

Introduction

Aseptic transport of axenic mice between laboratories requires a specially designed sterile shipping container. The shipping container is essentially a small isolator without forced ventilation. It usually consists of a flexible polypropylene container with disposable cages inside; two HEPA filters, one on top and the other on one end; and a disk seal through which materials or animals are moved between the container and the isolator. During shipping, the container is protected in a transport box. We supply food and water during shipping either with moistened chow or with commercial, sterile nutrient gel packets.

We use germ-free shipping containers purchased from Taconic (Figure 13.1; see Appendix). The process of moving mice between the shipping container and the isolator is similar to other types of internal transfer of mice (Chapter 11). The isolator and container are connected by a transport sleeve, the sleeve is sterilized, the mice are transferred, and the isolator and transport container are both sealed prior to removal of the sleeve. One difference is that the sleeve on the shipping container that we use measures 12" in diameter. The port on the donor isolator measures 18", necessitating use of a step-down transfer sleeve that is 18" at one end and 12" at the other end (see Chapter 3). Prior to attaching the transfer sleeve to the shipping container, we must connect the 12" shipping container sleeve to a freestanding 12" stainless steel port. This port will act as a coupling for the two connections.

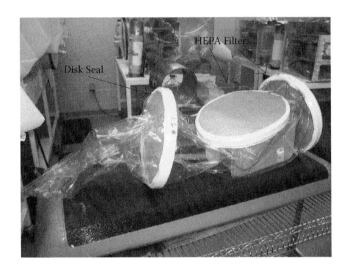

Figure 13.1 Shipping container.

Note: When shipping, there are two important steps in addition to the steps for performing a routine internal transfer:

1. Autoclave the stainless steel 12" port.
2. Soak the 12"-to-18" transfer sleeve for 60 minutes.

Procedure

1. Remove the shipping container from the shipping box (Figure 13.2). Place it on a clean surface. Look for the ethylene oxide (ETO) verification strip. The ETO verification strip confirms that the container was ETO sterilized.

2. Prepare the 12" port of the shipping container (Figure 13.3) by washing with sterilant. *Be careful.* The disk that seals the entire sterile shipping container consists of cardboard, Mylar® film, and nylon tape (Figure 13.4). If the Mylar is compromised, the cardboard disk may become wet and lose its structural integrity. This will cause a breach in the seal, resulting in contamination of the interior of the shipping container.

3. Prepare the autoclaved 12" stainless steel port by washing it with sterilant. Use the opportunity to introduce any other materials (water, caging, etc.) needed for the donor isolator as well. Wash and soak the 12"-to-18" transfer sleeve (Figure 13.5).

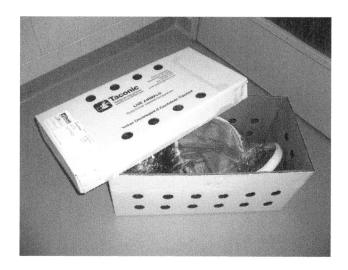

Figure 13.2 Shipping container in its box.

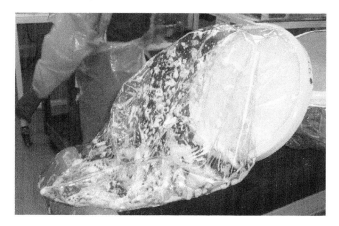

Figure 13.3 Prepare the sleeve of the shipping container.

4. Prepare the isolator port (see Chapter 7) and attach the 18"-to-12" adapter sleeve (Figure 13.6). With sterilant, wipe and wash a level surface for the 12" stainless steel port. The surface should be as clean as possible.

5. Attach the 12" end of the transfer sleeve to the stainless steel port (Figure 13.7). Use a flat (not grooved) 12" gasket. Place a 14" stainless steel clamp over the gasket.

6. Hand tighten the clamp. Make sure the 12" gasket is as straight as possible. As the clamp tightens, the gasket might slip. Make sure the clamp is tightened well, but not too tight.

Figure 13.4 Filter disk from shipping container. (For demonstration only: *Do not* remove the disk or get it wet.)

Figure 13.5 Cold sterilize the port and transfer sleeve.

7. While the 12″ stainless steel port is still wet with sterilant, take the 12″, open end of the shipping container and place it over the port. Dry the outside of the edge with alcohol and a dry, autoclaved terrycloth towel. Tape the edge with nylon tape (Figure 13.8). Keep the tape flat and tight.

8. Use nylon tape to cover/seal the edge of the shipping container port. Cover each previous edge with half the width of

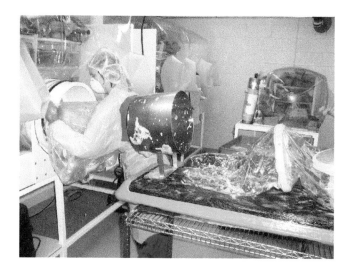

Figure 13.6 Place the transfer sleeve on the isolator port.

Figure 13.7 Attach the 12" port to the small end of the transfer sleeve.

the nylon tape. Tape back approximately 4". Once 4" have been covered, tape back over the edges of the tape (Figure 13.9). Once you have reached the starting point, cut the tape.

9. Using the atomizer, carefully inflate the sleeve and shipping container. Do not apply high pressure. Fog the interior cavity for about 40 to 60 seconds. As the shipping container inflates, guide the connections (e.g., the Mylar-covered disk seal) to their natural layout (Figure 13.10). Physically support the

Figure 13.8 Attach the sleeve of the shipping container to the 12" port.

Figure 13.9 Tape the shipper sleeve to the port.

noninflated section of the shipping container (far right) until it is fully inflated. Allow at least 4 hours of contact time.

Note: There is no way to reach inside the shipping container. You must manipulate the disk seal and cages within the shipping container from the outside. Use 60% isopropyl alcohol to lubricate the

Figure 13.10 Fog the transfer sleeve assembly.

Figure 13.11 Displace the disk to establish a passage between the donor isolator and the shipper.

exterior polyurethane. This will help your hands guide the objects without friction.

10. After observing the appropriate contact time, peel the nylon tape off the disk seal and carefully lay the disk Mylar side down (Figure 13.11). *Do not let the disk seal become wet.* There is a dotted line centered where the disk seal is seated when the shipping isolator is assembled. This dotted line will be used as a guide when you reseal the disk. Use isopropyl alcohol to lubricate the outside surface of the shipping container.

11. Manipulate and prop the disk seal above the cages.

12. Gently guide the cages down the sleeve into the donor isolator (Figure 13.12). Inside the transport isolator there is a strip of cardboard that is used to place on top of the cages. Leave the strip of cardboard in the shipping container.

13. Once the cages are in the isolator, place the mice in the cages with fresh bedding and moistened chow. Moisten chow by adding equal parts water and feed or use chow from the mold

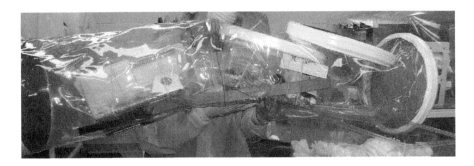

Figure 13.12 Move the cages into the isolator, leaving the cardboard strip inside the shipper.

Figure 13.13 Attach the cage cards to the cages.

trap (Chapter 9). Use the supplied rubber bands to attach the identification cards to the cages (Figure 13.13).

14. Once the mice are placed into the port, guide the cages back into the transport isolator the same way you guided them into the donor isolator.

15. Place the disk back into the original position. Use the dotted line marked on the container as a guide. Use pressure while taping the disk in place (Figure 13.14). *Notice the large marginal overlap over the edge of the internal disk seal.*

16. Carefully observe the cages and the tops. *Make sure the lids are on.*

Figure 13.14 When the labeled cages with mice are back in the shipper, replace the disk and tape firmly.

Figure 13.15 The reassembled shipper with mice is placed back into the box and secured for shipping.

17. Place the strip of cardboard inside the shipper on top of the cages. This will prevent the lids of the cages from direct contact with the polypropylene surface. *This step is very important.* If the lids are obstructed, the mice may suffocate.

18. Detach the shipping container, 12" port, and the 12"-to-18" transfer sleeve. Roll up the open end of the shipping container and secure the excess material with nylon tape.

19. Place the shipping container inside the shipping box and secure with plastic ties. Immediately label the box with the time and date (Figure 13.15).

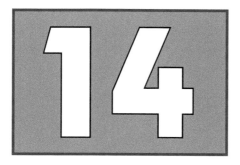

Rederivation

Introduction

Derivation (or rederivation) refers to the process of creating germ-free animals from stock that is not germ-free. The first germ-free animals were derived by cesarean section and were hand raised. Derivation of the first stable colonies of germ-free mice and rats was achieved and described by the Laboratory of Bacteriology at the University of Notre Dame (LOBUND; see Chapter 1).[1] Large animals such as dogs, pigs, calves, and others are still derived by cesarean section,[2-4] and species such as dogs, ferrets, and cats must still be hand fed. Today, however, propagation of germ-free rodents is usually achieved by maintaining axenic colonies that are bred in germ-free isolators, obviating the necessity for surgery and care of neonates. Breeding colonies can supply only those strains that are already germ-free, however. If new germ-free strains are required, they must be rederived into the germ-free state starting with breeding stock that is not germ-free. The recent increasing demand for new germ-free strains, including both inbred strains and genetically engineered mutant strains, has resulted in more frequent demand for rederivation of mice. Germ-free mouse strains can be derived in two ways: by cesarean section and fostering or by embryo transfer.

Derivation by Cesarean Section

Cesarean section is the traditional method of derivation. This method requires precise timing of pregnancies. Donor and foster dams must be date mated to ensure the availability of nursing germ-free dams for fostering, and donor mouse pregnancies must be timed and carefully monitored to ensure that offspring are viable and able to survive the stress of surgical delivery, transfer into the isolators, and fostering.

When parturition is imminent, the donor mouse is euthanized by cervical dislocation, and the pregnant uterus is transferred into the isolator by means of a "dunk tank" (Figure 14.1). The chamber of the tank has a divider that extends partway into the interior of the chamber, which is filled with 10% bleach. One end of the tank is aseptically attached to the recipient isolator via the port. The other end is protected by a cover, which is removed at the time of surgery.

Details of the procedure are as follows:

1. Breeding pairs of both the non–germ-free donor strain and the germ-free foster strain are date mated to ensure that parturition will occur simultaneously.

    a. It is best to start with as many foster pairs as possible because date mating in the isolators can be difficult, and not all dams will be appropriate fosters. Any strain can be used, but we prefer Swiss-Websters because they are good breeders and make good foster mothers.

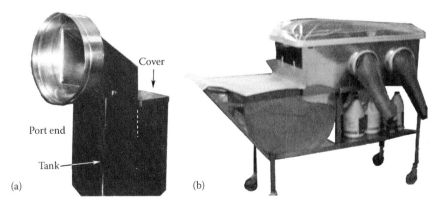

Cover

Port end

Tank

(a)   (b)

Figure 14.1 (a) Mouse dunk tank indicating the port end, tank cover, and position of the internal partial divider. (b) Large animal dunk tank for comparison. The pregnant uterus is transferred through the sterilizing bath directly into a self-contained isolator.

b. Foster breeders should be housed in a separate isolator from the rest of the colony in case of isolator contamination during transfer of pups.

c. Fostering is most likely to be successful if the donor and foster litters are of the same age, but the donor litter can be somewhat younger, as long as the foster dam is nursing.

d. Timing of the donor pregnancy is more critical than synchronizing timing between donors and fosters. This is because pups are mature enough to survive transfer and fostering only a few hours before parturition.

e. The best method of timing pregnancy is to pair one or two females with a male late in the day and to check for vaginal plugs every morning thereafter until mating occurs.

f. The length of pregnancy from breeding, as indicated by a plug, and parturition varies somewhat depending on mouse strain and implantation time. For this reason, it is wise to start with as many donor pairs as possible.

2. Pups must be derived as soon as possible before natural parturition begins. This maximizes their viability while ensuring aseptic transfer.

3. In addition to timed mating, pending parturition can sometimes be detected by palpation of the pubic symphysis of the pregnant dam. A slight relaxation of the ligaments can sometimes be detected.

4. Derivation of pups and transfer into the isolator must occur extremely rapidly to maximize pup viability.

a. Prior to derivation, the dunk tank must be in place and the surgical table and instruments sterilized and placed next to the open end of the tank.

b. At the time of parturition, the donor dam is euthanized by cervical dislocation, and the pregnant uterus is immediately removed. It is important to work quickly while maintaining aseptic technique.

c. The pregnant uterus is placed in a mesh bag and pushed down into the dunk tank below the partial divider. At the same time, a second person, working in the isolator, reaches into the tank and pulls the bag into the isolator. It is helpful to have a small metal hook for this purpose.

d. The uterus is immediately pulled from the bag, and the pups are removed. Sterile cotton swabs are used to gently dry the pups and stimulate them to breathe.

e. The pups are placed in the cage with the foster dam.

f. Usually, some or all of the foster's pups are removed and euthanized. It can be helpful to leave at least a few pups to encourage the foster to accept the new litter.

Derivation by cesarean section is fraught with potential difficulties, ranging from failure of date mating, premature or late derivation with loss of pups, failure of fostering, and contamination by microorganisms either by vertical transfer from the donor dam or by contamination during the transfer procedure. For this reason, many facilities prefer rederivation by embryo transfer. Embryo transfer is the method that we prefer.

Derivation by Embryo Transfer

Currently, there are two established methods of embryo transfer in mice: surgical[5,6] and nonsurgical.[7,8] In both cases, recipient germ-free females are paired with vasectomized germ-free males and monitored for the presence of vaginal plugs. At the appropriate time after mating (depending on the method of embryo transfer), females are aseptically removed from the isolators into a sterile biosafety cabinet (see Chapter 11), embryos are transferred, and the females are aseptically placed back into the isolator until parturition.

The surgical methods for vasectomy and embryo transfer are well described in the literature[5,6,9,10] and are not discussed here. However, there are some important differences between standard embryo transfer and germ-free embryo transfer. The most important difference is that, in the germ-free laboratory, whether transfer is performed surgically or via cervical injection, the process must be performed with *complete asepsis*.

1. This means that all the equipment, including the surgical hood, instruments, microscope, heating pad, and everything else, must be sterile. Microscopes, pipettors, slide warmers (used as heating pads during anesthetic recovery), and other electronics are ethylene oxide (ETO) sterilized (Figure 14.2; see Chapter 5).

Figure 14.2 Germ-free rederivaton surgery requires complete sterility of all materials and surfaces, including the hood, instruments, and surgeon (a), as well as the assistant (b).

2. Instruments, cages, and other equipment are autoclaved, and surfaces are cold sterilized.

3. The surgeon must use surgical personal protective equipment (PPE), and unlike conventional embryo transfer, a pipettor (rather than mouth pipetting) must be used (Figure 14.3).

At our institution, germ-free derivation surgeries are performed by the staff of the transgenic mouse core, obviating the necessity for our staff to become comfortable with the surgical procedures. *If this approach is used, however, it is important to stress to the transgenic core staff that complete asepsis is essential.* Unlike conventional animals, for which asepsis is only necessary for prevention

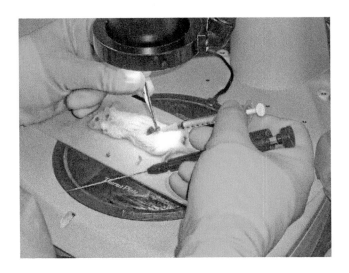

Figure 14.3 Sterile mouse transfer surgery. All materials, including the microscope and pipettor, must be sterile.

of infection and minimal contamination with nonpathogens is not a concern, germ-free animals must be kept germ-free. This means that absolutely no contamination is tolerable.

In addition to strict asepsis, there are other differences between germ-free and conventional embryo transfer. Germ-free animal colonies are generally small, cage space is at a premium, and the number of potential recipients is limited. As described for surgical rederivation, recipient dams must be kept in a separate isolator until parturition in case of contamination during the surgical procedure. Vasectomized males must also be kept in a separate isolator. Our practice is to vasectomize 10–20 males at a time at least 3 weeks prior to the planned derivation. This allows time to ensure that their isolator remains bacteriologically sterile. Recipient females can be placed in the isolator for mating, but after surgery, females are placed in a separate isolator to prevent contamination of the males. Because every surgical procedure is accompanied by a finite chance of contamination, however, the more recipients that are placed in the isolator, the greater the chance of contamination. For this reason, only two or three surgeries are performed at once.

Because of the small number of surgeries that can be performed at one time, it is important to maximize the chance that pregnancy will occur. This requires a reliable source of large numbers of embryos or blastocysts. For this reason, we prefer to use frozen embryos that are collected in advance and stored to ensure that a sufficient number

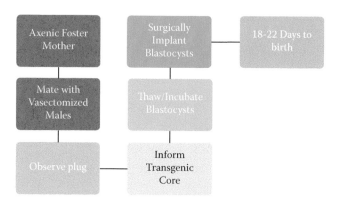

Figure 14.4 Our process for setting up germ-free embryo transfer. Collaboration with our transgenic core and use of frozen blastocysts improve our efficiency.

will be transferred. We have also found that while nonsurgical transfer diminishes the chance of contamination, it also requires more embryos to ensure implantation. For this reason, we prefer surgical embryo transfer. Figure 14.4 is a diagram summarizing our procedure for embryo transfer derivation.

Successful embryo transfer surgery will result in enough offspring to form the nidus of a new breeding colony (Figure 14.5).

Figure 14.5 Newly derived germ-free transgenic pups ready to be weaned.

References

1. Reyneirs J, Trexler PC, Ervin RF. Rearing germ-free albino rats. In: Reyniers JA, Ervin RF, Gordon HA, eds. *Lobund Reports.* Volume 1. Notre Dame, IN: University of Notre Dame Press, 1946:1–87.

2. Rohovsky MW, Griesemer RA, Wolfe LG. The germfree cat. *Laboratory Animal Care* 1966; 16:52–59.

3. Waxler GL, Schmidt DA, Whitehair CK. Technique for rearing gnotobiotic pigs. *American Journal of Veterinary Research* 1966; 27:300–307.

4. Griesemer RA, Gibson JP. The gnotobiotic dog. *Laboratory Animal Care* 1963; 13:SUPPL643–649.

5. Inzunza J, Midtvedt T, Fartoo M, Norin E, Osterlund E, Persson A-K, Ahrlund-Richter L. Germfree status of mice obtained by embryo transfer in an isolator environment. *Laboratory Animals* 2005; 39:421–427.

6. Okamoto M, Matsumoto T. Production of germfree mice by embryo transfer. *Experimental Animals* 2003; 48:59–62.

7. Cui L, Zhang Z, Sun F, Duan X, Wang M, Di K, Li X. Transcervical embryo transfer in mice. *Journal of the American Association for Laboratory Animal Science (JAALAS)* 2014; 53:228–231.

8. Bin Ali R, van der Ahé F, Braumuller TM, Pritchard C, Krimpenfort P, Berns A, Huijbers IJ. Improved pregnancy and birth rates with routine application of nonsurgical embryo transfer. *Transgenic Research* 2014; 23:691–695.

9. Van Keuren ML, Saunders TL. Rederivation of transgenic and gene-targeted mice by embryo transfer. *Transgenic Research* 2004; 13:363–371.

10. Suzuki H, Yorozu K, Watanabe T, Nakura M, Adachi J. Rederivation of mice by means of in vitro fertilization and embryo transfer. *Experimental Animals* 1996; 45:33–38.

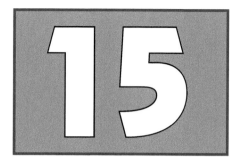

Microbiological Testing

Introduction

Frequent biological testing is essential for early detection of isolator contamination by microorganisms. Taxa that are of concern in laboratory animal colonies include protozoa, fungi, bacteria, and viruses. In germ-free colonies, because of the strong biological barrier, contamination by viruses and protozoa is less likely than in non–germ-free animals, and although we screen for these (albeit rarely), we have not experienced contamination. In our facility, the potential contaminants of more concern are bacteria and fungi.

Screening tests that we use in our facility and minimum frequency are as follows:

Fungi: For fungal contamination, we use a "mold trap" (see Chapter 9). Molds and fungi that may be slow growing or difficult to culture are detected by daily examination of the container of wet food that is placed in every isolator. Water is added as necessary to keep the mold trap from drying. All our isolators have mold traps, which are examined visually daily and by Gram stain on a weekly basis (see further discussion in this section). Mold traps can be helpful in detecting bacterial contamination as well. We do not culture specifically for fungi, although that is an option chosen by some laboratories.

Bacteria: Bacteria are the most common contaminants and therefore of the most concern. We have found that weekly culture of mouse feces and of the mold trap is the most sensitive method for detecting contamination (unpublished data). We do weekly cultures for aerobic and aerotolerant bacteria, and on a monthly basis, we culture a sentinel mouse for strict anaerobes.

In addition to culture, we examine Gram-stained smears of feces on a weekly basis. Although fairly heavy colonization is necessary to be detectable by Gram stain,[1] Gram staining can corroborate positive cultures and may detect bacterial species that cannot be cultured on artificial media. Theoretically, it will also detect fungi and protozoa, but in our experience, fungal contamination is rarely heavy enough to detect with Gram stains, and we have never experienced contamination with protozoa.

We have found that polymerase chain reaction (PCR) for detection of bacterial DNA in feces does not improve our detection of bacterial pathogens.[1] In our experience, all bacterial contaminants that were detectable by PCR were also detectable by culture, Gram stain, or both. For this reason, we currently use PCR only for verification of questionable results or for taxonomic identification of a bacterial contaminant. We have had the best results with quantitative PCR (qPCR) to detect a conserved region in the bacterial 16S RNA gene (see "Quantitative PCR" on page 196). Other laboratories describe similar methods.[2,3]

Suspicious or positive cultures are repeated immediately to verify the result. In addition, any isolator that produces a suspicious or positive culture or Gram stain is examined by qPCR and culture for strict anaerobes.

Serologic Screening: As noted, mouse pathogens (viruses, fungi, and protozoa) are not of great concern in the closed germ-free mouse colony. Nevertheless, for administrative reasons, we perform routine serologic screening quarterly. For this, one mouse in each isolator is removed and euthanized, and serum is collected for screening by our institutional rodent health surveillance service for serologic evidence of mouse hepatitis virus, mouse adenovirus, mouse parvovirus, minute virus of mice, endemic diarrhea of infant mice (EDIM), ecromelia, Sendai virus, lymphocytic choriomeningitis virus, polyomavirus, pneumocystis, and carbacillus. When mice are shipped, receiving institutions sometimes request additional screening.

Part 1: Screening for Bacterial Contamination

Supplies

- 1- and 2-mL autoclaved sample tubes
- Trypticase™ soy agar with 5% sheep's blood (TSA II™) culture plates
- 37°C incubator
- GasPak™ jars or similar airtight containers and anaerobe container sachets
- Glass microscope slides
- Gram stain kit
- Sterile swabs
- Butane torch
- Record log

Sample Collection

To verify the sterility of isolators, mouse feces are the most likely to reveal contaminants. For routine weekly sampling, it is best to sample feces from several cages in the isolator for Gram stain and culture and also sample the mold trap and any suspicious material for culture. We routinely culture for aerobic and aerotolerant bacteria. For detection of strict anaerobes, sampling of a sentinel mouse is necessary. Other potential sampling sites for detection of bacterial contamination are fur, water bottles, bedding, and swabs of isolator surfaces.

Fecal samples are collected directly from the anus, placed in screw-cap vials, and labeled. Vials are then removed through the port as previously described (Chapter 7).

Bacterial Culture and Slide Preparation

We use 5% sheep's blood agar plates incubated at 37°C either in air or in a sealed GasPak container system. The procedure is as follows:

1. Place the sample vials in a biosafety cabinet or flow hood (Figure 15.1).
2. Record the sample information on the culture log sheet (Figure 15.2; see Chapter 18).

Figure 15.1 Sterile sample vials are placed in a biosafety cabinet for culture.

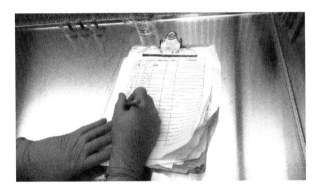

Figure 15.2 Record sample data on the culture log.

3. Label two blood agar plates per sample. Plates should be labeled on the bottom with an indelible marker (Figure 15.3).

4. Handle each sample separately. Use sterile cotton swabs to streak out the sample. Open the sample vial and aseptically remove a swab from the peel pouch. Use the swab to transfer a small amount of sample to the surface of each plate (Figure 15.4). You can use the same swab for both plates. Spread the sample over the surface of the plate.

5. Place one plate in the incubator and one in the anaerobic container. Plates should be incubated upside down. Proceed to the next sample.

6. After the plates are aseptically inoculated, prepare the slides for Gram stains. Prepare one slide for each sample as follows:

   a. Glass microscope slides with frosted ends are easiest to label.

Figure 15.3 Label two blood agar plates for each sample.

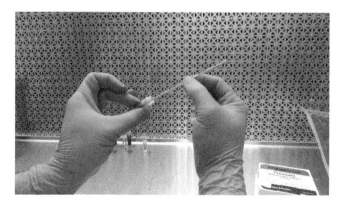

Figure 15.4 Use a sterile swab to aseptically collect a sample from the vial.

    b. Label the frosted end with the sample name and date (Figure 15.5). Use pencil because marker or pen will be washed off during the staining process.

    c. Spread a thin layer of sample over the surface of the slide. The material should be barely visible.

    d. Gram stain as described below ("Gram Stain Procedure," p. 195).

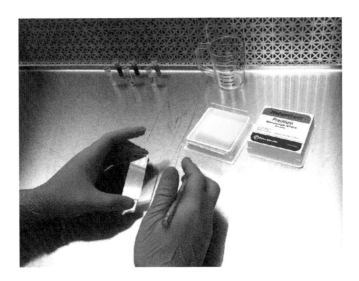

Figure 15.5 Label the frosted end of the slide with pencil.

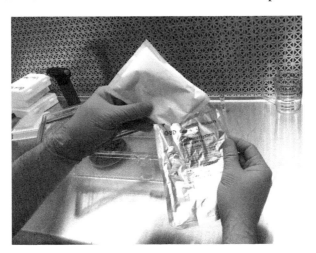

Figure 15.6 Open the GasPak and, if necessary, activate it according to the manufacturer's instructions.

7. Repeat the process for all samples using a fresh swab for each sample.

8. When all samples are streaked on plates and all slides prepared, activate the GasPak according to the manufacturer's instructions (Figure 15.6), place it in the anaerobic chamber with the plates, and seal the chamber (Figure 15.7).

9. Incubate both sets of plates at 37°C.

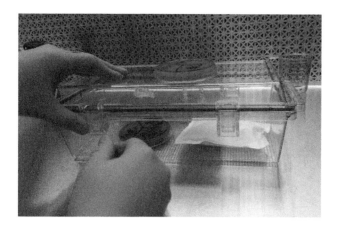

Figure 15.7 Place one plate from each sample in the GasPak chamber and seal the chamber. Incubate all the plates at 37°C.

10. Examine plates after 24 hours. If there is no growth, incubate for an additional 24 hours and reexamine.

11. Record the number and description of colonies.

12. Gram stain the colonies:

    a. If individual colonies are present, use a sterile swab or loop to sample one colony of each morphology, and prepare a glass slide as described in step 4, with the following difference.

    b. For staining of cultured bacteria, it is preferable to place a drop of water on the slide and suspend a small amount bacteria in the water (drop should be very slightly turbid). Air dry the slide before heat fixing and staining, as described in the next section.

    c. If a lawn is present, sample the edge of the lawn and stain.

13. We do not usually identify the bacterial contaminants beyond Gram stain and culture. However, in the case of monocolonization, it may be helpful to identify the contaminant regarding genus, if not species. This can be done by PCR amplification of the 16S ribosomal RNA gene or by routine microbiological methods (which are beyond the scope of this manual).

Gram Stain Procedure

1. It is helpful to use microscope slides that are frosted at one end.

2. Label the slide *with pencil* (ink will wash off during staining).

3. Prepare slides as described in the preceding section by spreading a thin layer of sample to be tested on the center of the slide.

4. Allow slides to air dry.

5. Heat fix by passing the slide *briefly* through an open flame. Do not overheat. The slide should be warm but not hot to the touch.

6. Use the Gram stain kit to stain according to the manufacturer's instructions.

7. Do not overdecolorize.

8. Allow slides to dry and examine as described in part 2.

Quantitative PCR

As noted, when PCR verification of contamination is needed, we use qPCR. We find that it is more accurate than gel-based detection methods. Detailed methods are available elsewhere,[1-3] and the following is only a summary of our procedure:

1. Collect at least one fecal pellet per isolator. You can use the same sample that you used for culture and Gram stain if sufficient material is available.

2. Isolate bacterial DNA with a QIAamp DNA Stool Mini Kit according to the manufacturer's instructions. Some laboratories prefer to use a bead beater to release bacterial DNA prior to extraction, but we have not found that necessary.

3. The PCR protocol is as follows:

   a. Applied Bioscience AB 7500 Fast Real-Time PCR System in 25 µL total volume

   b. Primers:

      i. Forward: CGATGCAACGCGAAGAACCT

      ii. Reverse: CCGGACCGCTGGCAACAAA

   c. Cycle times: 95°C for 10 minutes followed by 40 cycles of 95°C for 10 seconds then 55°C for 30 seconds

4. Germ-free and no-template controls are included in each run.

Part 2: Examination of Gram-Stained Slides for the Presence of Bacteria

Detection of bacteria by examination of Gram-stained slides requires practice and experience, both in use of the microscope and in recognition of bacteria. That said, in our experience a properly performed Gram stain is an extremely reliable way to detect bacterial contamination rapidly and inexpensively. The Gram stain distinguishes Gram-positive bacteria, which retain the Gram's crystal violet and stain blue, from Gram-negative bacteria, which do not retain crystal violet and stain with the counterstain (pink). Most animal and plant cells and other debris in fecal matter do not retain crystal violet and also appear pink, which makes it more difficult to identify Gram-negative bacteria (which can look like nonstaining debris) than to identify Gram-positive bacteria. Important considerations for successful use of Gram staining as a screening tool involve both the stain protocol and the examination and interpretation.

Tips for staining include guarding against the following staining errors, which can lead to false-positive or false-negative results:

1. Differentiation of Gram-positive and Gram-negative bacteria depends largely on the decolorization step. Over- or underdecolorizing can lead to spurious or inconclusive results.

2. Overheating during fixation can alter the morphology or destroy bacteria.

3. Most Gram stain kits include safranin as a counterstain. Fuchsin is an alternative that is somewhat brighter pink and renders Gram-negative bacteria somewhat more obvious.

4. A cheek swab is an excellent positive and negative control sample (see Figure 15.8).

The following are tips for successful examination and interpretation:

1. Slides do not need to be cover slipped but should be examined with a ×100 objective under oil.

2. Bacteria are recognizable by Gram-staining properties, size, shape, and homogeneity.

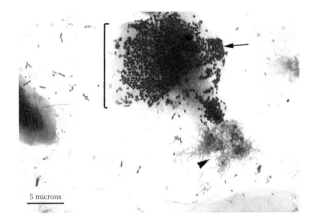

Figure 15.8 A cheek swab is a good positive Gram stain control. Mouths contain both Gram-positive (arrow) and Gram-negative (arrowhead) bacteria of a variety of shapes and sizes.

3. Bacteria must be distinguished from artifacts such as fragments of plant material from food, intestinal epithelial cells, or other debris.

4. Small numbers of bacteria cannot be distinguished from debris. There must be sufficient numbers of bacteria to allow recognition of the homogeneous shape and size.

5. Fortunately, most contamination events result in heavy colonization, resulting in the presence of many bacteria in the sample and easy detection in stained smears.

6. Other artifacts that may be confusing include plant or animal cell nuclei (may be confused with protozoa or yeasts) and precipitated stain.

The photomicrographs in Figures 15.8 to 15.16 illustrate the appearance of bacteria and of various nonbacterial artifacts in Gram-stained slides of mouse feces. Figure 15.8 is a Gram-stained cheek swab, which is a good positive control for both Gram-positive and Gram-negative bacteria. Figures 15.9 to 15.13 demonstrate the appearance of bacteria of various morphologies in Gram-stained fecal smears from monocolonized gnotobiotic mice. In general, bacterial density is high in these samples and easily distinguished from fecal debris based on the uniform size and shape of bacteria compared to background material. Larger bacteria (Figures 15.10 and 15.12) and Gram-positive bacteria (Figure 15.10) are more

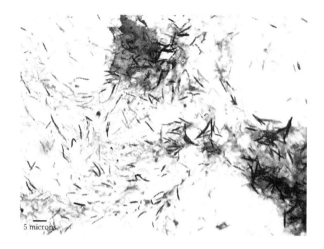

Figure 15.9 Gram-stained smear of feces from an SPF mouse. Many Gram-positive and Gram-negative bacteria are present.

Figure 15.10 Gram-stained smear of feces from a gnotobiotic mouse colonized with *Clostridium* sp. Many large Gram-positive rods are present.

easily identified than smaller, Gram-negative bacteria (Figures 15.11 and 15.13). Figures 15.14 and 15.15 demonstrate the appearance of common artifacts that must be distinguished from microorganisms. Figure 15.16 demonstrates the appearance of Gram-stained feces from accidentally contaminated mice. In our experience, accidental contamination of isolators results in rapid growth of bacteria, which reach easily detectable levels within 1–2 days.

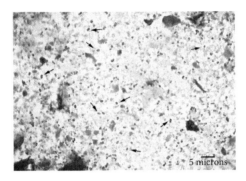

Figure 15.11 Gram-stained smear of feces from a gnotobiotic mouse colonized with *E. coli*. Short Gram-negative rods are similar to fecal material in staining properties but can be distinguished by their size and shape (arrows).

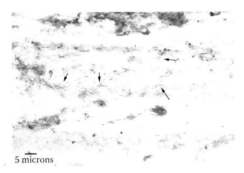

Figure 15.12 Gram-stained smear of feces from a gnotobiotic mouse colonized with *Bacteroides* sp. Long, filamentous, Gram-negative rods are evident (arrows).

Figure 15.13 Gram-stained smear of feces from a gnotobiotic mouse colonized with *Prevotella* sp. Small Gram-negative coccobacilli are distinct from fecal debris by their uniform size and shape (arrows).

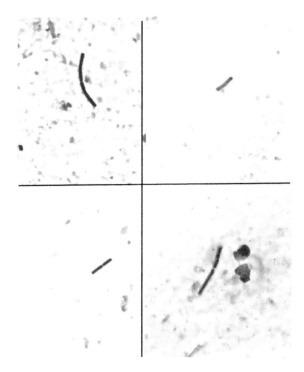

Figure 15.14 Gram-stained smears of feces from four germ-free mice. Individual bacteria-like particles are commonly present in low numbers in feces from germ-free mice. These particles may represent dead bacteria or plant debris from food or bedding and must be distinguished from living bacterial contaminants.

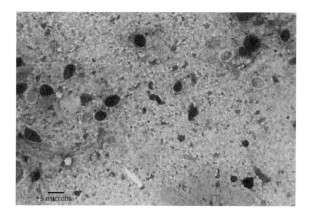

Figure 15.15 Gram-stained smear of feces from a germ-free mouse. Round-to-oval structures bear some resemblance to protozoa. They are likely plant cell nuclei derived from the diet.

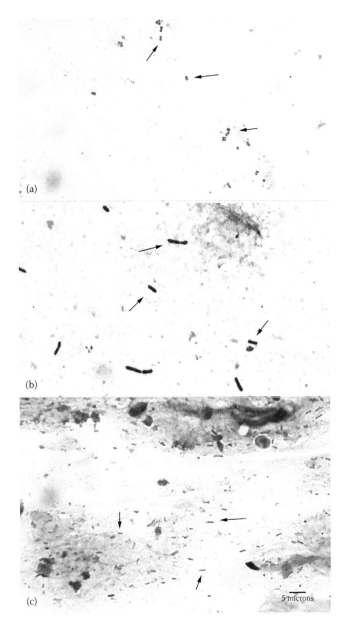

Figure 15.16 Gram-stained smears of feces from accidentally contaminated mice. Common contaminants are Gram-positive cocci such as staphylococci (a) and large Gram-positive rods such as *Bacillus* sp. (b), presumably from workers' hands and inadequately autoclaved food, respectively. Gram-negative rods (b) are less-common contaminants. Arrows indicate the bacteria.

References

1. Fontaine CA, Skorupski AM, Vowles CJ, Anderson NE, Poe SA, Eaton KA. How free of germs is germ-free? Detection of bacterial contamination in a germ-free mouse unit. *Gut Microbes*; 2015 May 27:1–9. [Epub ahead of print]

2. Arvidsson C, Hallén A, Bäckhed F. Generating and analyzing germ-free mice. *Current Protocols in Mouse Biology* 2012; 2:307–316.

3. Packey CD, Shanahan MT, Manick S, Bower MA, Ellermann M, Tonkonogy SL, Carroll IM, Sartor RB. Molecular detection of bacterial contamination in gnotobiotic rodent units. *Gut Microbes* 2013; 4:361–370.

16

Genetic Testing

Genetic contamination of mouse strains is an important concern for mouse breeding colonies in general, and a comprehensive discussion is beyond the scope of this manual. However, because germ-free mouse colonies are small and housing space is highly constrained, breeding germ-free mice presents some unique challenges that are worth discussion. Genetic contamination can be of roughly two types: genetic drift and inadvertent crossbreeding. Genetic drift occurs when mouse lines are propagated for many generations.[1-3] In large breeding colonies, it can be monitored and managed. In germ-free colonies, however, because of the limited number of breeding pairs available and the limited resources of most small operations, it is largely ignored. It likely does occur and should be a consideration in choosing control strains for experimentation, but it is not further considered here. Further information is available in the literature[2,3] and online (http://jaxservices.jax.org/genome/index.html, http://jaxmice.jax.org/genetichealth/GQCprogram.html).

A larger concern for germ-free colonies is inadvertent crossbreeding of strains. This occurs when, for example, a mouse escapes from its cage and is mistakenly returned to the wrong cage, when a technician is distracted and mixes up labels on cages, or when cage labels are lost or illegible. These occurrences must be guarded against but are unfortunately more common than one would expect, even in well-managed colonies. Germ-free colonies have unique challenges regarding potential genetic contamination. The major problem is the competing concerns of genetic and bacterial contamination. Because the chance of contamination by microorganisms is always present, every germ-free strain must be housed in at least two separate isolators (we prefer three if space is available). That way, if an isolator is contaminated and all mice must be culled, the entire strain is not lost. Because of this necessity, in smaller facilities each isolator must house more than one strain. The more strains there are in the isolator, of course, the greater the opportunity for inadvertent mixing of strains. Compounding the issue, germ-free breeding colonies tend to be small with few breeding pairs. Thus, every pair is valuable, and if genetic or bacterial contamination occurs, loss of even a single pair can place a strain on the colony.

The principal approach to prevention of genetic contamination is, of course, management. Training and establishment of standard operating procedures are essential for working with small colonies in limited housing. Consistent records (see Chapter 17) are necessary to prevent misidentification of cages. As far as is possible, strains in the same isolators should be of different coat colors, minimizing the possibility of error on the part of technicians. All breeding mice should be identified permanently with an ear tag. Escaped mice must never be returned to a cage but routinely culled to prevent errors in cage assignment. Cages with uninterpretable labels should be removed from the colony. Technicians should make a practice of double-checking cage identifiers when moving mice or changing cages and err on the side of caution if a mistake is suspected.

That said, mistakes happen, and in a small breeding colony they can be disastrous. For that reason, in our colony we genotype every breeding mouse. Ideally, mice should undergo a complete genomic screen. However, because our resources are currently limited, our practice is as follows:

- Replacement breeders are selected and ear tagged; at the same time, tissue is collected for genotyping. Mice can be selected prior to weaning but must be large enough to be permanently ear tagged.
- DNA is isolated and polymerase chain reaction (PCR) is performed to detect specific genetic mutations as follows:
  - Control mice (those with no engineered mutations) are screened for all mutations that are present within the same isolator. For example, if C57BL/6 mice are housed in an isolator with RAG KO mice, all C57BL/6 potential breeders are screened to ensure that the RAG KO construct is not present. RAG KO breeders in that isolator are screened to ensure that they are homozygous for the construct.

- If an isolator houses two mutant strains (e.g., IL-10 KO and RAG KO) as well as C57BL/6, all three strains will be screened for both constructs.

In some cases, a shortcut is possible. Many genetically engineered mutant mice carry a neomycin resistance cassette, which is retained in the genome following construction of the mutation. If this is the case, detection of the cassette in a mouse expected to be wild type can indicate cross contamination, regardless of the specific mutation.

If evidence of crossbreeding is detected, additional tissue is collected, and the screen is repeated. If it is still positive, the isolator must be cleared. The next steps depend on the status of the isolator and the value of the strains. If possible, the simplest solution is to cull all mice in that isolator and restock with mice of known, verified genotype. If there is a valuable strain involved, then it may be necessary to genotype all mice in the isolator and breed to retain or eliminate the desired or undesirable mutation.

Our protocol for PCR detection of mutations is as follows:

1. DNA isolated from the tail tip is preferred, but in older mice, ear-punch samples are easier to obtain.
2. DNA is isolated via routine methods using a Qiagen DNeasy Blood and Tissue Kit according to the manufacturer's instructions.
3. PCR reactions are run in 25-μL volumes on a Bio-Rad MJ mini thermocycler. Details of the cycles and timing are optimized for the specific mutation.
4. The protocol we use for detection of the neomycin cassette is as follows:
   a. Forward primer: 5'-TTC GGC TAT GAC TGG GCA CAA CAG-3'
   b. Reverse primer: 5'-TAC TTT CTC GGC AGG AGC AAG GTG-3'
   c. 2.5 mM Mg$^+$, 0.2 mM dNTPs (deoxynucleotide triphosphates), 0.5 mM each primer
   d. PCR cycles:
      i. Start: 95°C for 4 minutes
      ii. 30 cycles of 95°C for 15 seconds, 72°C for 30 seconds
      iii. End: 72°C for 4 minutes

5. Samples are separated on a 1.5% agarose gel at 100 V and 100 mAmp for 1.5–2 hours.

6. The presence of the neocassette is indicated by a 282-base-pair band.

References

1. Petkov PM, Cassell MA, Sargent EE, Donnelly CJ, Robinson P, Crew V, Asquith S, Haar RV, Wiles MV. Development of a SNP genotyping panel for genetic monitoring of the laboratory mouse. *Genomics* 2004; 83:902–911.

2. Proetzel G, Wiles MV, eds. *Mouse Models for Drug Discovery.* New York: Springer, 2010.

3. Casellas J. Inbred mouse strains and genetic stability: a review. *Animal* 2010; 5:1–7.

17

Record Keeping

Introduction

Consistent and reliable records are essential for any research enterprise, and this is particularly true for the germ-free laboratory. Guarding against microbial contamination and genetic contamination, keeping track of experimental and animal use protocols, animal health, ordering supplies, and maintenance of equipment are all dependent on good record keeping. Details, of course, will vary according to the needs of each facility. What follows is an overview of the major types of records we have found useful for husbandry and experimental procedures.

Our permanent records are all computerized. Separate documents are kept for each category of records, and one technician is assigned to ensure that the computerized records are kept up to date and accessible to the lab. Larger facilities may have to divide this task between two or more people. In addition to computerized records, we keep written records to keep track of the daily status of rooms, isolators, hoods, and equipment (see the discussion that follows). The written records are posted on the room doors or equipment as appropriate, and the information is transferred to the computer as often as necessary. When the written record is filled, it is put in a binder or folder for storage.

We divide our records into three categories:

I. Colony management documents

   A. Population census

Colony Management Documents

Population Census

The population census log (Figure 17.1) is used to document all mice in the breeding colony, including breeding pairs, replacement breeders, and offspring that are not yet assigned to an experiment. It contains all of the information from each cage card and is updated weekly. We find it most useful to cross reference cage numbers by position of the cage on the rack in each isolator. This minimizes the risk of cage misidentification, which could lead to mixing of strains and genetic contamination (see Chapter 16). As an additional precaution, all breeding mice are ear tagged prior to placing in breeding pairs or harems, and the mouse tag number is recorded on the census sheet.

Communication between technicians after moving or weaning cages during husbandry or health checks is important. Notes should be made on a daily basis to keep information on new litters, mouse requests, and newly weaned mice up to date and easily accessible.

Breeding Records

All potential breeders are genotyped and ear tagged prior to mating. Our breeding mice are housed in either pairs or harems of one male and two females. Each breeding cage has an individual breeding record, which records all information for that pair or harem (Figure 17.2). Information about the pair or harem, including strain,

Germ Free Mouse Population Census

Isolator	Location on Rack	Strain	Breeding Cage (X = Yes)	# of Females	# of Males	Date of birth	Ear Tag number	Litter DOB	# of Pups in Litter	Date Tag & Tailed	Notes
	1										
	2										
	3										
	4										
	5										
	6										
	7										
	8										
	9										
	10										
	11										
	12										
	13										
	14										
	15										
	16										
	17										
	18										
	19										
	20										
	21										
	22										
	23										
	24										
	25										
	26										
	27										
	28										
	29										
	30										

Figure 17.1 The population census log is a summary of the cages in each isolator and their status. More detailed information for breeding cages is recorded on the breeding record (see Figure 17.2).

Germ Free Breeding Record

Strain:_____

Mated:_____

Isolator:_____

Breeder Info	Male	Female	Female
Ear Tag #			
Date of Birth			
Date Tailed			
Notes			

Genetic Monitoring	Male	Female	Female

Litter #	DOB	# Females	Female Ear Tag #	# Males	Male Ear Tag #	# of pups	Notes
1							
2							
3							
4							
5							
6							
7							
8							
9							
10							
11							
12							
13							
14							
Retire							

Notes and Cage Activity

Cage Deactivation Date:

Figure 17.2 An individual breeding record is kept for each breeding pair or harem.

age, ear tag, date of birth, and genotyping results, as well as litter information, is kept on the breeding record. Litter birth dates and the number of pups in each are updated at least once a week. Recording the number of pups produced by each female and the frequency of the litters helps to determine when to cull or retire a breeding cage.

Genetic Monitoring

Organized and thorough records that track genetic monitoring (see Chapter 16) are crucial to maintaining a germ-free breeding colony. Diligence is required to prevent genetic contamination as different strains of mice are generally cohoused in large breeding isolators. When DNA samples are collected from mice, the isolator, ear tag number, strain, and date—at minimum—must be recorded. It is also helpful to make copies of the original genotyping submission forms, if any. When genotyping results are received, we record them on the cage card in the isolator, on the breeding records, and on the genetic monitoring log (Figure 17.3).

Isolator Activity Logs

In addition to the individual breeding records, population census, and genotyping records, we keep a running log for each breeding and experimental isolator (Figure 17.4). The activity sheets record bacteriologic monitoring, health monitoring, husbandry activities, entries and exits, repairs, and other maintenance information. When appropriate, similar activity logs may be kept for biosafety cabinets and Isocage racks.

Project-Planning Documents

Germ Free Mouse User Request Form

Our facility functions as a core, providing mice and services to researchers across campus. Because we are a relatively small facility in a large institution, we always have a waiting list for mice, housing, or both. The principal purpose of the Germ Free Mouse User Request Form (Figure 17.5) is to help us keep track of who needs mice; what kind of mice they need (strain, sex, and age); what kind of housing they need; and how long the experiment will last.

Genetic Monitoring Log

Isolator	Cage	Date Tagged	Ear Tag Number	Sex	Strain	Genotyping results	Results Date	Notes

Figure 17.3 Replacement breeding stock are genotyped and ear tagged, and the results are recorded on the genetic monitoring log.

This information is essential for planning and ensuring fair distribution of limited resources. In addition to planning, the Germ Free Mouse User Request Form helps ensure that animal protocols are approved and that appropriate facilities are available to perform the requested procedures. The Germ Free Mouse User Request Form provides the germ-free technicians with information on the strain, age, number, and sex of mice; desired start date; and experimental space required for the project. Also included is a general overview of the experiment and a sample collection schedule. It is helpful to be able to quickly reference the experimental schedule, so post it near the experiment.

Experimental Isolator Activity Sheet

Date	Health Checks	Isolator Pressure	Change Water	Change Bedding	Gm stain Culture	Entered Waters	Port Entry	Comments/ Supplies Entered

Figure 17.4 Activity logs are kept for breeder and experimental isolators.

Mouse Project Requests

As noted, we are a heavily used facility and almost always have a waiting list for mice. Usually, we are able to fill requests in a timely fashion, and we generally work on a first-come, first-served basis. To help this process run smoothly, in addition to the Germ Free Mouse User Request Form that provides details of each request, we keep a running list of requests for mice and space (Figure 17.6). This helps us plan ahead and distribute mice equitably among users. When a Germ Free Mouse User Request Form is received, the request is recorded with the lab or institution's name, the primary contact, date

ULAM
Unit for Laboratory Animal Medicine

Germ Free Mouse User Request Form

Date of request:____________ Requested start date:____________

Contact Information
PI:
Short code:
Approved Protocol #:
Protocol Exp Date:
Main Contact (MC):
MC Phone#:
MC Alternative Phone#:
Email Address:

Mice Requested and Summary of Experimental Procedures			
Strain and sex of mice requested		Number of groups	
Number of mice per group		Total number of mice	
Infectious agents to be administered		Route of administration	
Other materials to be administered		Route of administration	
Duration of experiment		Preferred age of mice requested	

Experimental Space Requirements and Sample Storage			
Experimental space requirements	☐ Tecniplast caging system ☐ Isolator ☐ Bio safety cabinet (class II)	Sample Storage	☐ 4C ☐ -20C

Protocol Description of procedure(s)/Service(s)
This area may be copied and pasted directly from the approved protocol. Only include information that pertains to germ free mice. If you do no have access to an electronic copy of the protocol/amendment, Indicate the appropriate sections from within the document *Example: Section 11.E.8A Procedure Information, Injection section.*
Procedures

Drugs/Experimental Agents and Route (If Used)
Feces Collection or Body Weights Required (with Time Points or Frequency)
Hazardous Substances (If Used), including pathogenic organisms
Other Procedures to be performed (eg, urine, blood collection, etc). Please specify frequency of collection, and method of labeling, handing and storage of samples

Figure 17.5 The Germ Free Mouse User Request Form includes the number of mice requested, the housing requested, experimental details, and the expected duration of the experiment as well as animal use approval information.

Mouse Request Log

	Lab/ Institution	Contact	Date Requested	Strain	Number	Sex	Age	Estimated Date Available	Location of mice	Start or Ship Date
1										
2										
3										
4										
5										
6										
7										
8										
9										
10										
11										
12										
13										
14										
15										
16										
17										
18										
19										
20										
21										
22										
23										
24										
25										
26										
Projects On Hold										

Figure 17.6 The mouse request log is a summary of all current requests, date of request and when available, and the location of the mice to be assigned to the project.

requested, requested start date, and the strain, number, age and sex of the mice for the experiment. Recording isolator location and information about the mice on this document helps technicians keep track of mice that are on hold for experiments.

Experimental Space Availability

In our laboratory, experimental space includes small isolators, Isocages, and microisolators in biosafety cabinets. Like mice, housing space is often limited, and we keep a running log of experimental requests. The Germ Free Mouse User Request Form includes a check box for users to request specific types of housing. Those requests are recorded on the appropriate housing form (Figure 17.7) to be filled when space becomes available.

Other Documents

Bacterial Screening Log

All samples and results pertaining to the germ-free (axenic) or gnotobiotic status of the colony need to be easily accessible and clear to all members of the germ-free core. Every Gram stain and culture result is recorded on the screening log (Figure 17.8). When samples are collected by technicians, the samples are assigned an accession number and sent to the lab for further processing. This procedure has proven useful to highlight positive or unusual results to easily track contaminations or concerns.

Health Monitoring

Colony mice that are not assigned to experimental protocols are observed at least once daily for evidence of illness (for experimental mice, protocols for observation and treatment are dictated by the animal use protocol). Should an abnormality be noted, it is recorded in the health log (Figure 17.9), and the mouse is monitored as appropriate. If the injury is minor, the mouse is observed until the issue is resolved. Major illnesses or injuries are rare, and if they occur, the mouse is usually euthanized. Mice that die or are euthanized due to illness are necropsied and the results recorded in a necropsy log.

Isocage availability

Rack 2 - Side 1 Availability

	A	B	C	d	E	F
6	Magnetic: not available					
5						
4						
3						
2						
1						

Rack 2 - Side 2 Availability

	G	H	I	J	K	L
6						
5						
4						
3						
2						
1						

Bio-Safety Cabinet Availability

Hood Number	Lab/Institution Current Project	Start Date	Date Available	Next Project	Start Date	Date Available
1						
2						
3						
4						
5						
6						
7						
8						

Pending Projects: Hood

Hood Needed	Current Project	Desired Start Date	Estimated Date Available	Hood Number	Hood Number

Experimental Isolator Availability

Isolator	Current Project	Status	Date Available	Next Project	Start Date	Date Avail.
Exp. 1						
Exp. 2						
Exp. 3						
Exp. 4						
Exp. 5						

Figure 17.7 The space available forms for isolators, Isocages, and biosafety cabinets facilitate efficient assignment of mice to experiments.

Bacterial Screening Log

Date		Source of Sample	Accession #	Test Results				
Submitted	Completed	Isolator, Tecniplast Rack/Side, Hood #		Gram Stain	Anaerobic Culture	Aerobic Culture	PCR	Other

Figure 17.8 The bacterial screening log is essential for ensuring sterility and tracking contamination when it occurs.

Paper Documents

As noted, the computerized documents are updated regularly based on written records that are kept in the animal rooms. These documents include room logs, isolator logs, hood logs, and culture logs (Figure 17.10).

Germ Free Animal Health Reporting Log 2015

Date	Room #	Isolator	Strain	Tag	BOD	Animal Health Concern	Action Taken	Resolution (monitor, cull, resolved)

Figure 17.9 The health reporting log ensures that mice are treated or euthanized in a timely manner.

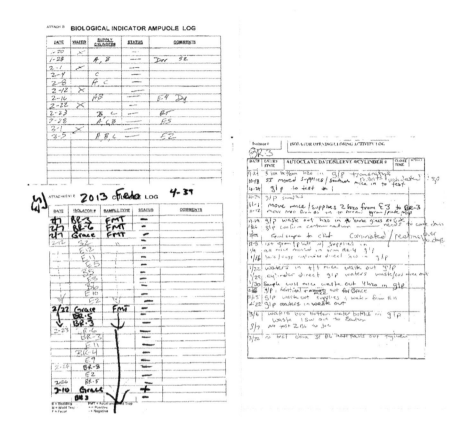

Figure 17.10 Examples of paper records used for daily recording of culture results and other activities. Paper records are kept in the animal rooms so that they are easily accessible. Results are transferred to the computerized forms as appropriate.

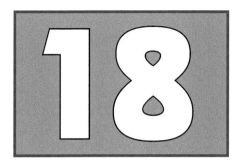

Facility Setup and Management

Introduction

Every facility is unique. The size of the facility, its overall goals, sources of support, and other factors will in large measure determine the details of how the facility is organized and run. That said, there are a few principles to be considered and decisions to be made prior to establishing a new facility. In this chapter, we will discuss the decisions to be made in establishing a new facility, such as the considerations of staffing and personnel management. Our suggestions are based on our own experience establishing our facility, as well as on our experience in assisting other institutions in establishing their own facilities.

Establishing a New Facility

In our experience, there are two major truisms regarding establishment and management of gnotobiotic facilities. The first is that running a gnotobiotic laboratory is labor intensive, fraught with potential complications, and extremely expensive. The second (related) truism is that it is not possible to fully fund a germ-free facility from user fees alone. External support from the institution or grant support in association with a program project or other large grant or contract is essential for success.

There are several categories of decisions that must be addressed in the planning of a new facility:

1. Funding sources
2. Space requirements
3. Isolators, equipment, and other housing options
4. Personnel
5. Fee structure

These are addressed in the following sections.

Funding

In today's research environment, few laboratories have access to unlimited funds for core support. To our knowledge, as of this writing there is only one germ-free facility in the United States that is supported directly by federal funds. The University of North Carolina has a gnotobiotic rodent resource center (http://www.med.unc.edu/ngrrc) that is supported at least in part by the National Institutes of Health. This mechanism of support is no longer available to other institutions, however, and the University of North Carolina's resources are available to outside investigators on only a limited basis, if at all. For most institutions, nonfederal support is necessary.

As noted, gnotobiology is expensive. Labor costs are high because basic husbandry requires many staff hours not only to prepare and assemble isolators and equipment but also to provide a constant supply of feed, water, bedding, clean cages, and other supplies. Experimental procedures that are relatively simple outside the isolators, such as fecal sample collection, ear tagging, and genotyping, are complicated and labor intensive when performed behind the germ-free barrier. In addition to labor, gnotobiology is space intensive. Square footage that could potentially house hundreds of specific pathogen free (SPF) mice can only house three or four isolators (depending on the isolator size) and perhaps as many as 50–100 mice. Finally, the equipment itself (isolators, biosafety cabinets, Isocages, etc.) is expensive, and even after the initial cost, replacement parts and supplies such as tape and filtration materials add to the cost. For these reasons, a reliable source of income is essential.

Potential sources of financial support for gnotobiotic laboratories may include user fees, institutional support (through either center grants or other internal sources) and core support from large multiuser grants or contracts. Because user fees cannot support the full cost, the first consideration for investigators planning to establish a new facility is to identify funding sources. It is wise to begin by establishing commitments for support from institutional funds as well as from any core grants or major potential users before initiating further planning. At the very least, establish your approximate costs for equipment and the first few years of salary support before taking any action.

Space Requirements

Germ-free laboratories require not only large amounts of space but also space with very specific characteristics. The animal rooms must be large enough to house isolators and attendant equipment with sufficient room for moving supplies and for working around the isolators without damaging them. Doors and hallways must be large enough to easily move large equipment in and out without damage. The electrical supply must be able to support heavy use by blower motors and biosafety cabinets, and emergency backup power is essential. There must be convenient access to well-functioning autoclaves of sufficient size to house supply cylinders and equipment carts, and technical support for autoclaves, cage-washing equipment, biosafety cabinets, and so on must be reliable and available in emergency situations. Access to ethylene oxide sterilization, dedicated laundry facilities, and distilled or reverse osmosis purified water is also required. In addition to gnotobiotic-specific requirements, basic resources for animal research such as high-temperature tunnel washers, small autoclaves, environmental controls (temperature monitoring of rooms, light times, etc.), and filtered air supply should be available.

Layout of the facility is also important. In addition to room size, placement and access may be issues. The rooms should be large enough to house equipment but separate enough that breeding isolators, which must be kept in a strictly clean environment, can be housed separately from experimental isolators, and experimental isolators housed separately from biosafety cabinets and Isocage racks. Rooms with germ-free animals must be kept completely separate from gnotobiotic animals and from experimental procedure space. Large storage spaces for autoclaved supplies and equipment must be conveniently placed but separated from animal rooms. Biohazard waste, if it is to be stored in the facility even for a short time, must be assigned separate space. Access to autoclaves and washing facilities must be convenient. If possible, non–germ-free procedure space and lab space should be nearby, and methods for transporting animals from isolators to laboratories or to non–germ-free housing should be established. If rederivation is planned, dedicated space for surgeries is helpful. A small room with a surgical hood and room for one or two small isolators would be ideal for this purpose. As a general approach, it is wise to discuss space requirements with institutional directors and establish commitments for adequate space based on the projected needs of the individual facility.

Isolators, Equipment, and Housing

Equipment needs will be based on projected use of the facility. Decide on the number, size, and type of isolators and other housing that you plan to use. Our facility uses large isolators for breeding, smaller isolators for experiments and supplies, and biosafety cabinets and Isocage racks for short-term housing. Smaller facilities and facilities with minimal need for experimental or gnotobiotic housing may need fewer or smaller isolators. Some facilities use small, stacked isolators for short-term experiments. Larger facilities may need more or larger isolators or even clean rooms. It is worthwhile to discuss potential needs with users as well as to investigate the types and sizes of equipment available and the space available to estimate projected startup costs.

In addition to major equipment, determine the ancillary equipment need, such as supply cylinders, carts, racks, replacement supplies, and so on. The lists provided in previous chapters may be helpful in determining the types and numbers of supplies needed.

Personnel

Because gnotobiology is an extremely labor-intensive activity, personnel are the backbone of the endeavor. Dedicated and well-trained staff are essential for the success of any facility. Staff not only must be dedicated to the success of the facility and well trained in the minute details that are required to maintain sterility, but also must be extremely observant and attentive to detail. The best germ-free technicians are sufficiently familiar with the physical characteristics of the animals and isolators that they can detect subtle changes that may presage difficulties. As noted in Chapter 15, for example, small alterations in odor or appearance of the isolators or animals can be the first indication of a bacterial contaminant. Similarly, small errors, such as inadvertent dropping of a fecal pellet or bedding frag-ment during cage changes must be noticed and corrected. Records must be kept meticulously (see Chapter 17), and animals and equip-ment must be examined at least daily for signs of illness or damage, respectively. If staff are involved in experimental procedures, the ability to follow protocols and understand the aims of the experiment are also essential. Thus, a principal goal in establishing and main-taining a gnotobiotic facility is identifying, training, and retaining outstanding staff.

Identifying Staff

Individuals who are already trained in gnotobiology are rare, and most facilities train their own staff. Recruiting individuals with training in laboratory animal husbandry can hasten learning and allow workers to become competent more quickly than if they start with no husbandry experience. Our practice, as much as possible, is to recruit new staff from the institutional laboratory animal care unit. These individuals already understand the basics of animal care, are experienced in observing laboratory animals for signs of illness, and may have laboratory animal technician certification or other training. They are comfortable with experimental protocols and recognize the importance of record keeping. Also, if recruited from within the institution, they are familiar with institution-specific rules and regulations.

Ideally, regardless of the size of a facility, a gnotobiotic laboratory should have at least two full-time workers. This is because working in the isolators often requires two people working together, and as noted in previous chapters, many of the manipulations performed either inside or outside the isolators are difficult or impossible for one person to perform alone. It is possible to run a small facility with one full-time and one part-time worker, but it is essential to have at least one dedicated person who is consistently present and an assistant who is available on call. Hiring full-time permanent staff ensures consistent oversight and adherence to standard protocols. Temporary or part-time staff, while helpful if assistance is needed, may have difficulty maintaining the consistent practices that are essential to preventing contamination.

The most important characteristic of a germ-free technician is dedication to the job. Running a germ-free facility is hard work, requires constant vigilance and attention to detail, and can be frus-trating and disappointing. Workers must be motivated to succeed, able to work through setbacks, and care deeply about the success of the enterprise. That said, several other characteristics are helpful:

1. A certain amount of physical strength to work with isolators and equipment
2. Willingness to work with laboratory animals
3. Sufficient "people skills" to work closely with investigators and their staff while safeguarding animal health and equipment condition and sterility
4. A basic understanding of and interest in scientific research, research practices, and protocols

Training

The specialized functions of a gnotobiotic animal laboratory are best learned by experience. We have found that adding staff to an established facility can be accomplished by on-the-job training, and that husbandry personnel familiar with the running of an animal facility can learn the protocols for gnotobiotic techniques when experienced individuals are present. When establishing a new facility, however, it is helpful to have one or two individuals visit a working laboratory for training. As of this writing, several facilities, including ours, are willing to host visitors and provide various levels of instruction and advice. An excellent resource is the newly established gnotobiology e-mail list, administered by the University of Alabama (https://listserv.uab.edu/scgi-bin/wa?A0=GNOTOBIOTICS).

Retention

Germ-free technical staff must be handled with care. Germ-free technicians are highly trained, dedicated professionals who bear complete responsibility for the success of the facility, including everything from quality control and animal health to research progress. A few important considerations for their care include

1. Pay them well, as far as is allowed by institutional regulations.
2. Treat them with respect: utilize their expertise and include them in decisions.
3. Do not micromanage.
4. Foster a culture of teamwork. Avoid hierarchies and allow each worker to take responsibility for his or her specific interest or specialty.

Fee Structure

The details of setting fees for institutional services are well beyond the scope of this manual and will likely be determined by each individual institution regardless of the needs of the core facility. That said, there are a few unique characteristics of gnotobiotic husbandry that should be clarified with the institutional administration to assist them in establishing user fees.

1. As noted, user fees cannot be expected to fully fund the facility. Therefore, fee structure should be established with the goal of setting a price that will contribute significantly to the upkeep of the facility while remaining within the budget of the investigators.

2. Setting per diem rates is complicated by the fact that gnotobiotic animals may be kept in a variety of different types of housing, and the housing unit for which fees are charged may differ.

   a. Mice can be housed in cages within isolators, in cages in biosafety cabinets, or in cages in Isocage racks. Each of these types of housing requires different husbandry manipulations.

   b. When mice are housed in isolators, the basic unit of charge could be either the isolator or the cage.

   c. If the isolator is the basic unit of charge, per diem rates could include the daily husbandry costs as well as setup and tear down of the isolators.

3. Based on these considerations, the institution must determine if there is to be a single per diem per cage (or per mouse) regardless of the type of housing, or alternatively, if each type of housing should be assigned a unique per diem rate. If the former, costs must be somehow averaged between housing types. If the latter, a decision must be made regarding whether fees are to be based on the cage or the isolator and how differences in husbandry will be taken into account.

4. Unlike SPF mice, germ-free mice are usually bred in-house rather than ordered from a vendor. For this reason, the institution must set a fee for mice as well as for per diem rates.

   a. This fee could be estimated based on the estimated cost of producing a mouse. That would require estimating the per diem cost of husbandry in the isolator, the daily census in the breeding colony, and the average number of pups per breeding cage per unit time.

b. Alternatively, the cost could be estimated based on a standard rate and adjusted according to the requirements of the individual institution.

5. In addition to mice and per diem rates, fees should be set for technician time and standard experimental procedures (fecal collection, body weights, blood and tissue collection, etc.). These can be based on similar fees for technical services provided elsewhere, but the added time and training involved in handling of gnotobiotic animals should be considered.

6. If rederivation or other surgical services are offered, they should be assigned a fee. As noted in Chapter 14, rederivation surgery in the germ-free facility is more complicated and difficult than similar procedures performed under SPF conditions. These difficulties should be considered when setting fees and also when estimating the time required to produce a new strain.

In summary, because of the unique requirements of germ-free husbandry and gnotobiotic research, and the impossibility of recouping the true cost of doing business from user fees alone, we find that the simplest approach to setting fee structures is the best. We charge the same per diem rate for all germ-free and gnotobiotic mice, regardless of housing type. We set individual fees for commonly performed procedures and as much as possible include the cost of labor and supplies in the unit fee for each procedure.

Appendix: Sources for Equipment and Supplies

Isolators, Isolator Supplies, and Housing

- CBClean, in Madison, Wisconsin, makes isolators (standard and custom sizes) and supplies most parts and supplies, including filter housing, sterilization cylinders and equipment, HEPA filters, and so on. If they do not supply something, they will probably know where to obtain it: http://www.cbclean.com.

- Harlan Laboratories, based in Indianapolis, Indiana, also makes isolators: http://www.harlan.com/.

- The Standard Safety Equipment Company (Chicago, IL) can supply custom-built isolator canopies and gloves: http://www.standardsafety.com/.

- Tecniplast (Milan, Italy) makes ISOcages and racks: http://www.tecniplast.it/en/index.html.

- Allentown, Incorporated, supplies many varieties of animal caging and housing systems: http://www.allentowninc.com.

- Taconic Biosciences (Germantown, NY) supplies shipping isolators: http://www.taconic.com/.

Other Equipment

- We get our atomizers from Spraying Systems (Wheaton, IL): http://www.spray.com/.
- Grainger, a supplier of tools and industrial equipment, also supplies isolator gloves, blower motors, and other similar supplies: http://www.grainger.com.
- Steris, based in Mentor, Ohio, supplies sterilization indicators and supplies as well as water bottles for autoclaving: http://www.steris.com/.

Sterilants

Most sterilants are available directly from the manufacturer:

- Spor-Klenz® (a mixture of peracetic acid and hydrogen peroxide) is available from VWR International: https://us.vwr.com/.
- Alcide Expor® (chlorine dioxide) is now owned by Ecolab: http://www.ecolab.com.
- Clidox® (chlorine dioxide) is manufactured by Pharmacal Research Laboratories: http://www.pharmacal.com.
- Steriplex SD® (a new sterilant of undisclosed formulation) is manufactured by Steriplex: http://www.steriplex.com/.

Other Supplies

- Many supplies are available from standard laboratory suppliers, such as Fisher Scientific (http://www.fishersci.com) and VWR International (https://us.vwr.com/).
- Mylar® polyester film may be purchased from Piedmont Plastics: http://www.piedmontplastics.com.
- High-temperature tape can be ordered from Grainger: http://www.grainger.com.
- Nylon tape can be ordered from VWR.
- Pesola scales (http://www.pesola-scales.com) can be ordered from the manufacturer or local suppliers.

Index

#0004 - 130317 - C254 - 229/152/14 - PB - 9781498736329